Programming

ARDUINO

WITH PYTHON FOR ROBOTS (2020 EDITION)

A Beginner to Advanced Reference Guide to Arduino Microcontroller Processing and Robotics

TED

HUMPHREY

Copyright

Contents

CHAPTER ONE

CREATING THE PROGRAMMING ENVIRONMENT FOR PYTHON AND ARDUINO

This chapter is essential if you are a beginner who doesn't understand much about python and creating an enabling environment for Arduino on your computer. It doesn't matter which category of operating system you are running on your computer as this chapter will take you through the python and Arduino programming environment.

Getting Started with Python

In 1991, Guido Van Rossum brought the python programming language which was at the time a basic language of instruction for learners and students of programming. Today, the python programming language is among the most used and user-friendly programming language. The programming language works on all platforms, such as Windows (all versions of Windows), Linux OS, and Apple OS. The python programming language, recently, has gone through some minor modifications when newer versions have been developed to replace the older versions. The switching between python v2 to python v3 was met with huge applause. One main difference between python v2 and the newer versions (python v3) is that the v3 can handle both English and Non-English characters while the v2 can only work with English characters. Also, the python v3 is a bit more compatible and smaller than the v2. Nonetheless, the python v2 is still being used to teach the basis of python programming today.

Installing python on your computer

As it has been said above, the python programming language has two versions which are the v2 and the v3 versions. Most operating systems, such as Linux distributions and Macs OS, have their default python which has been pre-installed on them. The python V2 is the default python on most operating systems. The python 2.7 is the last version of the v2.x series (x here represents unknown v2 series) while the python 3.8.5 is the latest version of the V 3.8 series as at the time of writing this guide. The python 3.8 will be used as a guide in this book because it is user-friendly and it is easy to understand the basics. The instructions in this guide will work on all versions of the python 2.x and the 3.8 series. The python's installation procedures for three of the most common operating systems will be discussed.

Linux distributions: Installing python on Linux distributions (such as Ubuntu and Fedora) is quite easy unlike the Windows operating systems. This is because the Linux distributions come default with python that has already been installed. Your job will be to check, using some sets of basic instructions, if you are running the right version of python. If you want to check the version of python that has been pre-installed on your computer, enter the command line $ python –V. If you execute the command line properly, you will get a window displaying the version of the python you are using.

Endeavor to use a capital letter V in the command prompt. If the version on your Linux OS is the latest version, that is good. You are on the path to getting started with python. However, if the version on your computer is below the version 3.8, you will have to upgrade the version to the V3.8 series.

If you are using the Ubuntu operating system, you should have a pre-installed python latest version. If you don't have the python latest installed, you can upgrade to the latest version by upgrading with the command line: $ sudo apt-get update && sudo apt-get --only-upgrade install python.

Installing python on Windows

Installing your python on the Window OS is not as simple as the other operating systems. The first thing is to go to http://www.python.org/getit to download the python version 3.8. which is the latest version of python To get it right, navigate to the system property of your computer to confirm the type of operating system you are running. You will be able to see whether you are running the 32bits OS or the 64bits OS. Once the python file has been downloaded successfully, you can then run the file for full installation. As soon as the installation is complete, you can explore the python command line tool and IDLE from the start menu. If you want to make the python accessible from the Windows command prompt, you need to set the path in the *environment variable*. The path is the most used environment variable in Windows is the **path.** Follow these steps to add the python path for easy execution;

- Scroll to the bottom pane of your Window interface, and search "**Advanced system settings**" in the "**type here to search**" box.
- The **"system properties"** window will be prompted which is usually open to the "**advanced tab**" by default. Scroll to the bottom and select **environmental variables.**

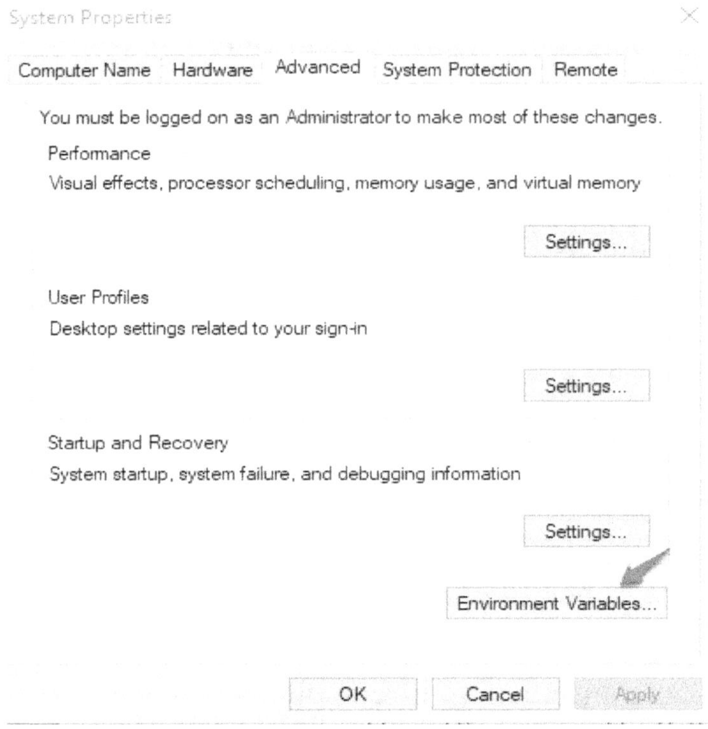

- From the **environmental variable** window, you will get two sets of variables: the system variable and the user variable. Both of the variables have the **path** variable and you can decide to edit any of the paths (either from user variable or the system variable). You can choose to edit the path for the user variable section if you are planning to be using the command for only your user account on the computer. You will choose the system variable if you need the change to be effective across all users sharing the computer with you.

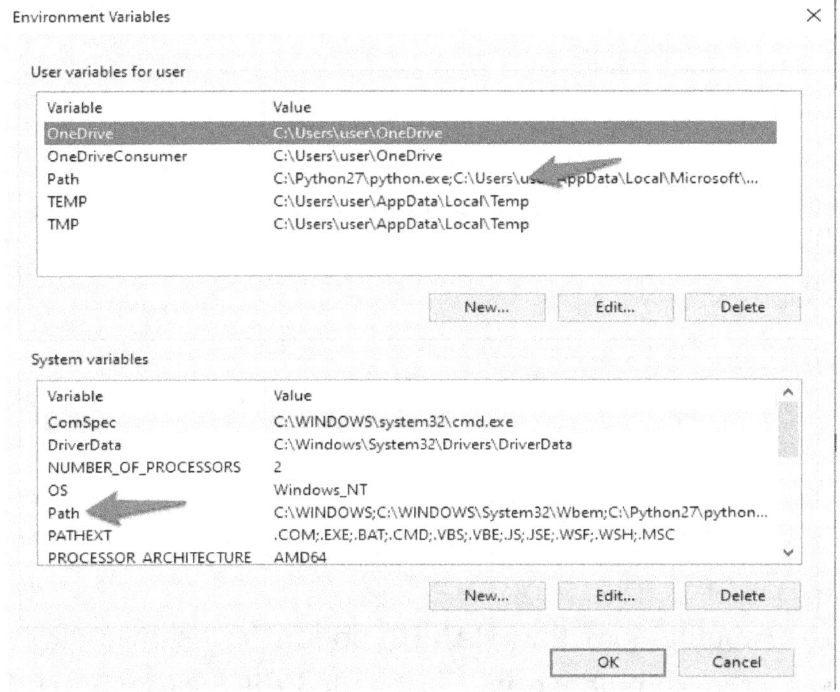

- Tap on *path* and then scroll to the right section to choose *edit.*
- Add a new path by simply tapping on **"New"** and a new line will be added at the bottom of the list where you can either type or paste the **path** of the python that you downloaded. You can also choose the **browse** option which will enable you to browse your files and copy the program file.

- Tapping on the **PATH** option as shown in the screenshot above will bring an **Edit System Variable** window. Add C:\Python27 or the full path of the custom Python installation directory at the end of the existing **PATH** variable. You should put a semicolon (;) right before the installation path.

Installing python on Mac OS: The Mac OS comes preinstalled with a copy of python, only that the rate of updates for the default version is not really fast. The version of Mac OS called the 10.9 Maverick comes default with python 2.7.5.

Installing python Setup tools

The python Setup tools is a library featuring an assemblage of utilities that users can use for building and distributing Python packages. The most used tool from this assemblage is called **easy_install**. It enables you to access the PyPI, the de facto python repository that features the highest number of Python packages is the PyPI (http://pypi.python.org). The PyPI also gives easy ways to install as many packages as possible on your operating system. With the PyPI, you will be able to install any package by name on your python. The *easy_install* utility will be able to automatically download, build, install, and manage your packages for you as a user. Many of the packages that will be used for the python and the Arduino project in this book will be installed by the **easy_install.** It is necessary to state here that the python experts all over the world have been, of recent, using another tool called **pip** to replace the **easy_install** utility. This is because the **easy_install** utility fails to carry out some basic tasks such as support for uninstalling packages, tracking actions and support for other versions control systems. The **easy_install** and **pip** utility use the same python repository; they can both be used to install any python package of your choice.

Let us get started by using the **easy_install** to install various python's setup tools for different operating systems;

For Linux operating system

-	In Ubuntu, the setup tools can be installed by prompting the command:

$$\textbf{\$ sudo apt-get install python-setuptools}$$

Note the use of the $ symbol in the prompt above which is the command prompt in the Linux based system. It is often used to specify that the python command you prompted should be run on a command line and rather not on a python/perl/ruby shell. The % symbol in front of the command line also means the same thing. You might not necessarily put the $ symbol in front of the command line as this will not destroy the command line.

- You can install the setup tools on Fedora by using the **yum** which is the Fedora's default software manager it can be installed using the default software manager yum:

$$\textbf{\$ sudo yum install python-setuptools}$$

- Most other Linux distributions can be downloaded and built when you used using the following single line command:

$$\textbf{\$ wget https://bitbucket.org/pypa/setuptools/raw/bootstrap/ez\_ setup.py -O -| sudo python}$$

Once you have successfully installed the setup tools on your Linux distribution, the easy_install can now be accessed directly right from the terminal as a built-in command.

For Windows operating systems

The procedures for installing the setup tools for Windows are not exactly as simple as it is for Linux based operating system. The installation needs you to get the right ez_setup.py file straight from https://pypi.python.org/pypi/setuptools. Once the file has been downloaded successfully, press the **shift** button on your computer and then right-click the actual folder containing the setup files. You can then prompt the command;

>>> **python ez_setup.py**

The command line above will be able to install python Setup tools in the Scripts folder of the default Python installation folder. You should adopt the same method you used when you wanted to add the python to **Environmental Variables.** Insert the setup tools by including C:\Python27\Scripts to **path.** This will allow you to be able to install various python packages using the **easy_install** inside the python packages folder named Libs. Close and then open the command prompt window after you have added the packet manager to your environment variable. This is to allow these changes to be effective.

For Mac OS

You can install the Setup tools in Mac OS by following the method below

1. Navigate to the Mac section https://pypi.python.org/pypi /setuptools to download the **ez_setup.py** file.

2. Open the terminal and browse to the directory where the file has been downloaded to. You can, most of the times, see the downloaded file inside the download folder on your computer.

3. Run the command below inside the terminal to allow you to build and manage the Set up tools.

$ sudo python ez_setup.py

Installing pip

The setup tools have been installed successfully, you can then use it to install the pip.
- To install the pip on Linux or Mac, try and run the command line **$ sudo easy_install pip** to install the pip.
- For Windows OS, access the command prompt window and then prompt the command line **> easy_install.exe pip**

If the pip has already been installed either by you or by default, you need to upgrade it to the latest edition. This is necessary to fix certain bugs that might be present in the version you are running. To upgrade pip, enter the command **$ sudo easy_install --upgrade pip**

Installing Python packages

Having successfully installed the pip, you can use two different to install any outside python package available on the PyPi repository at http://pypi.python.org. To work effectively with installation of python packages, you have to be familiar with the methods identified below. Note that the **PackageName** has been used in these examples to replace any third-party python package since there are many packages you can actually download from the python's repository. If you have a name of the exact third-party you want to download, simply replace the term **PackageName** with the name of the third-party app.

Note that there are instances where you will require superuser privilege (root privilege) before you will be allowed to install or uninstall any third-party app. Try and use **sudo** and then follow it by the appropriate command for the cases.

To install a Python package, prompt the following command at the command window:

$ easy_install PackageName

Or you can use the pip by prompting the:

$ pip install PackageName

Prompt the command below if you have a specific third-party ap in mind that you want to install;

$ easy_install "PackageName==version"

If you don't know the precise version number, try and use the comparison operators like >=, <=, > or < to include a range for the version number. When you include a range like this, the easy_install and the pip will be able to find and install the right version of the package from the python's repository and then install it for you. Use **$ easy_install "PackageName > version"**

You can use the following commands for pip to allow you to install the same package you installed with the **easy_install;**

$ pip install PackageName==version

$ pip install "PackageName>=version"

For instance, if you want to install a package version between 3.0 and 5.0, you can enter the command **$ pip install "PackageName>=0.3,<=0.5"**

Upgrading a package with either the **easy_install** tool or the **pip** utility is not difficult as both commands are even similar;

$ easy_install --upgrade PackageName

$ pip install --upgrade PackageName

The **easy_install** utility does not exactly favor clean package un-installation of a package; you can tweak this to your favor by making the program stop searching for the specific package. You can then, later, delete the package file from the directory where it has been installed:

$ easy_install -mxN PackageName

The pip even gives a good way to carry out clean uninstallation of packages, simply prompt;

$ pip uninstall PackageName

Getting started with the python basics

This section should be thoroughly revised by users who have little to no understanding of python basics. If you are a user that has already understood the python basics, you can move to the next chapter. Nonetheless, whether you are a beginner or a pro, this section will get you intimated with some crash concepts in python programming language. You have already installed python from the appropriate website, what is left now is to open the python IDLE from the location where it has been installed so that you can have access to the full python interface. The python IDLE means **integrated development environment.** The IDLE will give you a basic text environment (built-in text editor) that allows you to create and test python scripts easily. To have access to the python 3.8 IDLE, simply enter IDLE in the search bar at the bottom of the window. You will get the search result where you can tap to open the python IDLE interface. The python

IDLE has a development mode called the interactive mode which consists of many features needed to run and create python scripts effectively. The python 3.8 IDLE will look like the screenshot below;

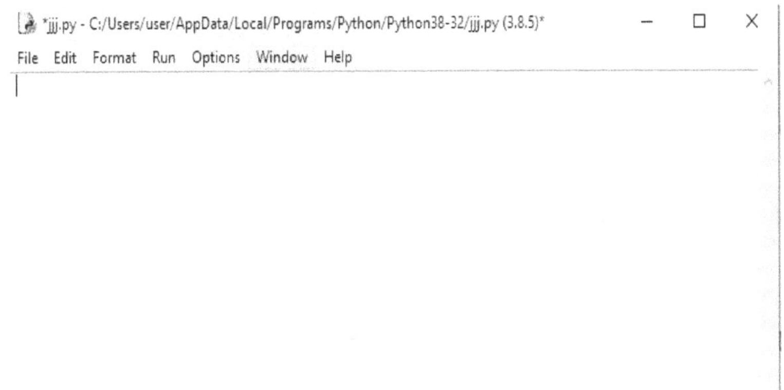

As you can see in the screenshot above, the python's interactive mode has menu features which are displayed at the upper section of the window. These features include;

- **The file menu:** This menu allows you to open new python scripts, save your result and access some files. You can even print the current window from the file menu.

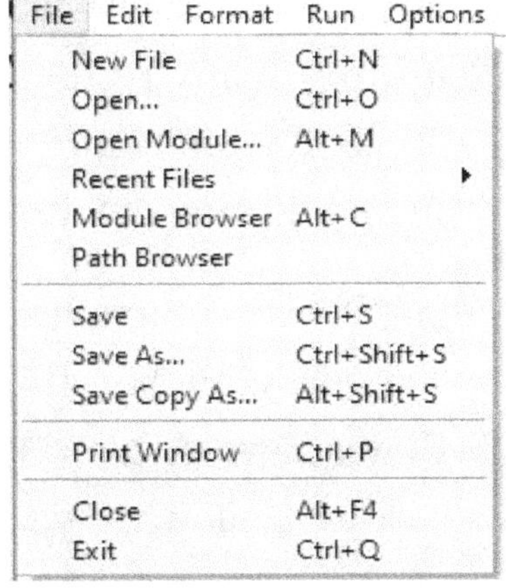

- **Edit menu:** Allows you to copy, paste and do a lot of other editing features.

Edit	Format	Run	Options	Window	Help
Undo					Ctrl+Z
Redo					Ctrl+Shift+Z
Cut					Ctrl+X
Copy					Ctrl+C
Paste					Ctrl+V
Select All					Ctrl+A
Find...					Ctrl+F
Find Again					Ctrl+G
Find Selection					Ctrl+F3
Find in Files...					Alt+F3
Replace...					Ctrl+H
Go to Line					Alt+G
Show Completions					Ctrl+space
Expand Word					Alt+/
Show Call Tip					Ctrl+backslash
Show Surrounding Parens					Ctrl+0

- **The Run menu:** This allows you to run your command. You can alternatively press the F5 key on your keyboard to run any script. Note that you will be asked to save your prompt into a directory before the script will run and display the answer.

Check the following "Do it yourself" instruction after you have successfully opened the python IDEL;
- In the IDLE 3 window, at the >>> prompt input print and then wait to look at what you have on the screen.

You will observe that the print command has been changed to a violet color. This is because the python language sees the print statement as a built-in function. The color aids easy recognition of the python's syntax and makes the scripts look beautiful and logical.

- Press the spacebar on your computer and input something like ("This is my computer") and then wait a little bit. You will observe that the statement (This is my computer) is colored green. This is because the python language sees the statement as a string literal. These are the ways python color your statement for more understanding and clarity.

Can you spot the color change in the screenshot below?

```
*jjj.py - C:/Users/user/AppData/Local/Programs/Python/Python38-32/jjj.py (3.8.5)*

File  Edit  Format  Run  Options  Window  Help

print ("this is my computer")
```

- You can close the python shell by pressing **CTRL + Q** on your computer or navigating to the file menu at the top section and scroll down to select "Exit." Another way you can use to close a python script or window is to prompt the command **exit ()** and then press the F5 key to run. You will get a window that looks like the one below;

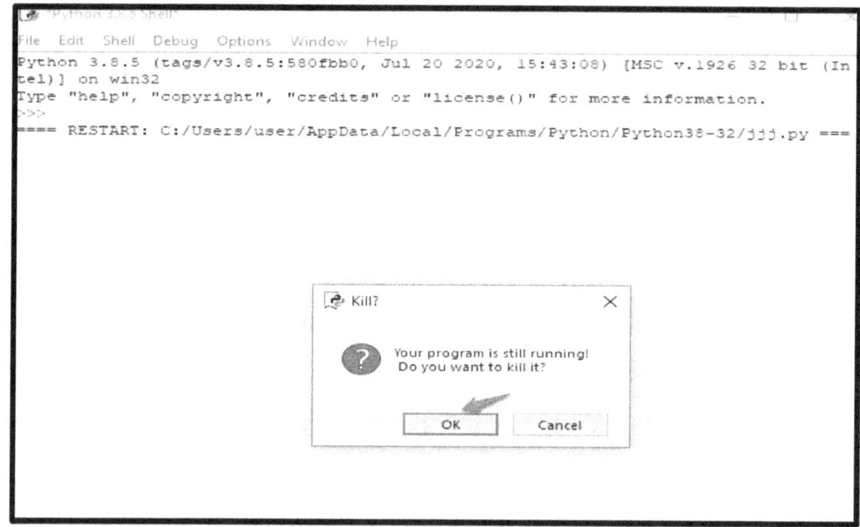

- Creating a Python script in IDLE is not really difficult. You start by opening the text editor (IDLE) window by pressing or by choosing the file menu at the upper part of the window and choose "New window". The new window will be opened with the "untitled" boldly written at the top section. This is the IDLE text editor. The python statement won't be showing results in the mode (untitled mode) even when you hit the "enter button". This is because the script has not been saved. To get past this limitation, enter your python statement and you will either be prompted to save it or you can choose the "save" option by yourself from the file menu. This will save the script to a file.

The print function in python

The usefulness of the print function is to show the result for a specific argument. The items you put inside the print function are called *argument.* The syntax of the print function is print (argument). The argument part of the print function can be characters like 123 or ABC. These characters are also known as string literals. To enter any character using the print function, you must enclose the character with either a single or a double quote ('or "). The print statement below uses single quote while the second on uses the double quote to enclose the characters;

>>>Print ('This is my name')

This is my name

>>>Print ("This is my name")

This is my name

Notice that the quotation mark did not accompany the print result. In the same vein, you should not mix the single quotation mark with the double quotation mark in one print statement; if you do, the screenshot below is what you will see

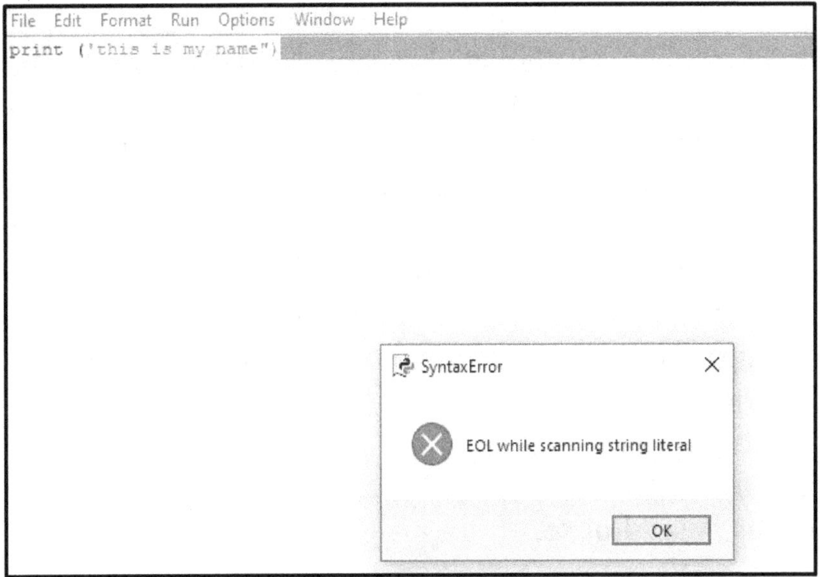

Protecting a single quote with a double quote: Sometimes, you might want to display a result that already has a single quotation mark inside showing possession, you should use a double quotation mark to enclose the result.

>>>print ("This is my father's house")

This is my father's house

Protecting a double quote with a single quote: If you have a double quote within the statement you want to print, you should try and use a single quote to enclose the print statement. See an instance below;

>>>print ('I said that, "I need a house!" and I bought one.')

I said that, "I need a house!" and I bought one.

Controlling your output with escape sequence in python

You can use an escape sequence in python to let a python statement jump or escape from the usual behavior or action. Escape sequences usually start with the backlash character (\). A typical instance of using the escape sequence to insert special formatting in output is the use of the **\n** escape sequence. The **\n** escape sequence brings any character you listed after it on to another line (next line). See below;

>>> **print ("This is my name.\nThis is my wife.\nAnd this is my house.")**
This is my name.
This is my wife.
And this is my house.
>>>

Breaking a long output lines of characters

The first method you can use to break a long output line of characters is to use the string concatenation. The string concatenation is able to take many strings of texts and combine them to have one string of text. The plus (+) is often used to achieve this. You will also use the escape sequence function to jump out of the normal print function (use the backlash \ for this purpose). Essentially, the two things that you need to concatenate are + and \.

```
>>> print ("I went to the market and bought vegetables, rice" +\
... "and a lot of fishes!")
I went to the market and bought vegetables, rice and a lot of fishes!
>>>
```

The version 3.8.5 might be having problems with the above string literals by displaying an error called EOL (end of strings literal), simply use the plus (+) sign only to concatenate. For example,

```
S= ("aaa" + "bbb")
      Print (S)
S = aaabbb
```

Creating comments in scripts

In scripts, comments are words or notes from the person (author) that compiled the Python script. The comment serves to give a clearer picture of the script's syntax and logic. The python interpreter doesn't particularly have anything to do with comments as it will ordinarily ignore it. However, comments are important ways to enable users to modify or debug a script. Precede a script with the hash symbol (#) to add a comment to the script. The python interpreter will not reckon with anything that comes after the hash sign. For instance, the comment can give you a way of adding your name, the time the script was written and the purpose of the script. You can put the comment at the top of the script or at the bottom of your script.

You can break up sections of the script by inserting long lines of the hash (#) symbol. This provides more clarity and separates the comment from the main body of script. In addition, you can also add comments at the end of the statement. For example, print () # insert a blank line in output.

Understanding the Python variables

You can see a variable as a name that keeps a value for future use in a script. A variable is just like a cup of tea. A teacup typically holds tea, of course! But a tea cup can also hold coffee, water, stones, rocks, gravel, sand etc.
While naming your python variable, you need to know that the variable names are case sensitive. For instance, the variable SandStone and sandstone are two different variables. Some rules that guide creating the names of variables in python include;
- Don't use a python keyword as a variable name. How can you know if a variable name is a python keyword? More on this later.
- Don't use a number as the first character of your variable name.
- Don't insert any spaces inside your variable name.

Python Keywords

The list of python keywords changes every now and then, therefore it is always handy to know the list of some current python keywords before you start creating your variable name. All the python's keywords except true, false and none are in lower case letters. Find below the list of python's keywords;

['False', 'None', 'True', 'and', 'as', 'assert', 'break', 'class', 'continue', 'def', 'del', 'elif', 'else', 'except', 'finally', 'for', 'from', 'global', 'if', 'import', 'in', 'is', 'lambda', 'nonlocal', 'not', 'or', 'pass', 'raise', 'return', 'try', 'while', 'with', 'yield']

Creating Python Variable Names

The first character in your python variable name must not be a number but can be any of the following; a letter from lower case "a" to lowercase "z", a letter from uppercase "A" to uppercase (Z) and the underscore (_). After the first character, you can use any of the following characters to precede it; the numbers 0 through 9, the letters a through z, the letters A through Z. and the underscore (_) character.

Since you cannot use a space in a variable name, you can replace it with an underscore (_). The variable name teacup can be tea_cup. You cannot use a variable without a value in python, it is therefore important you assign value to the variable name before you use it.

Assigning Value to Python Variables

You will be able to assign value to a variable by writing the name of the variable first, then put an equal to (=) symbol and then add the value you want to assign to the variable. The syntax is **Variable = value.** For instance,

>>>Tea_cup = "tea"
>>>print (tea_cup)
Tea

```
hgt.py - C:/Users/user/AppData/Local/Programs/Python/Python38-32/hgt.py (3.8.5)
File   Edit   Format   Run   Options   Window   Help
Tea_cup = "Tea"
print (Tea_cup)
```

```
Python 3.8.5 Shell                                               —    □    ×
File   Edit   Shell   Debug   Options   Window   Help
Python 3.8.5 (tags/v3.8.5:580fbb0, Jul 20 2020, 15:43:08) [MSC v.1926 32 bit (In
tel)] on win32
Type "help", "copyright", "credits" or "license()" for more information.
>>>
==== RESTART: C:/Users/user/AppData/Local/Programs/Python/Python38-32/hgt.py ===
Tea
>>>
```

Formatting Variable and String Output

When you use a variable, you automatically add extra formatting issues to the script. For instance, the print function will include a space anytime it sees a comma in the script. This is the reason you don't need to put a space at the end of the text that initially contains a comma. If you want to separate a string of character from the variable in the output, simply use a separator in the statement. The below code uses a separator (sep) to put an asterisk in the result instead of space;

>>> **print ("I enjoy my", tea_cup, "!", sep='*')**

I enjoy my*tea*!

```
>>> tea_cup = "tea"
>>> print ("I enjoy my", tea_cup, "!", sep='*')
I enjoy my*tea*!
>>>
```

Learning about python data type

When you create a variable by an assignment such as *variable = value*, Python will determine and assign a data type to that particular variable. A *data type* determines how the variable will be stored and the rules that govern how the data will be exploited. Python deploys the value it assigned to the variable to determine the type of the variable.

When the Python statement coffee_cup = 'tea' was prompted, Python observed the characters in quotation marks and classified the variable coffee_cup to be a *string literal* data type, or str.

You can actually determine the category of data type Python has given to a variable by deploying the type function. See the example below where the variable coffee_cup is a string literal variable and the variable cups_consumed is an integer variable.

>>> **coffee_cup = 'coffee'**

>>> **type (coffee_cup)**

<class 'str'>

```
>>> cups_consumed = 3
>>> type (cups_consumed)
<class 'int'>
>>>
```

Python assigned the string literal (str) data type to the variable coffee_cup since it saw a string of characters between quotation marks. But the python language saw the whole number for the cups_consumed variable, hence it was assigned the integer data type.

The print function will assign the string literal data type to its argument. This is the case for anything that is in the form of an argument, such as numbers, quoted characters, variable values etc. This means that you can mix different data class in your print function argument. The print function will change all the data type present to a string literal data type.

If you make a little change in the cups_consumed variable will make python change the data type. For instance, if you change the number given to cups_consumed – say from 3 to 3.2 – the python will reassign the data type it formerly assigned to cups_consumed from integer to float (of course now we have a decimal point)

```
>>> cups_consumed = 3
>>> type (cups_consumed)
<class 'int'>
>>> cups_consumed = 3.2
>>> type (cups_consumed)
<class 'float'>
>>>
```

Allowing Python Script Input

There will be a particular that you (script writer) will need your script users to enter some data into your script from their keyboard, you can use the python's input function to do this. The input function is a python's in-built function and deploys the syntax below;

variable = input (user prompt)

In the script below, the variable cups_consumed is assigned that was provided by the script user who answered the question "How many cups did you drink? The script writer provided the prompt (which is meant to be answered) to the script user in the form of argument. The user will provide the answer and hit the enter key. This will make the input function to assign 3 as the answer to the variable cups_consumed.

```
>>> cups_consumed = input("How many cups did you drink? ")
How many cups did you drink? 3
>>> print ("You drank", cups_consumed, "cups!")
You drank 3 cups!
>>>
```

For the user prompt, you can enclose the string of the prompt with either a single or a double quote.

As a script writer, do well by adding a space at the end of each prompt to allow the users to be able to answer the prompt without difficulty. It is not good typing in an answer that is jammed with the prompt. The input function considers all input as strings which are actually different from the way

Python handles other variable assignments. Remember that if the variable cups_consumed = 3 were inside your Python script, it would be given the integer data type. When using the input function, as shown below, the data type is assigned to string, str.

>>> **cups_consumed = 3**
>>> **type (cups_consumed)**
<class 'int'>
>>> **cups_consumed = input("How many cups did you drink? ")**
How many cups did you drink? **3**
>>> **type (cups_consumed)**
<class 'str'>
>>>

You can use the integer function to change which were input from the keyboard from strings data type. The integer function will change a number from a string data type to an integer data type. You can as well use the float function to change a number from a string to a floating-point data type. The script below shows how to change the variable cups_consumed to an integer data type.

>>> **cups_consumed = input ("How many cups did you drink? ")**
How many cups did you drink? **3**
>>> **type (cups_consumed)**
<class 'str'>
>>> **cups_consumed = int(cups_consumed)**
>>> **type (cups_consumed)**
<class 'int'>

>>>

You can decide to do some little manipulations here and use a nested function. The *Nested functions* are functions within another function. The general format of a nested function is as follows:

variable = functionA(functionB(user_prompt))
The script below deploys this method to effectively change the input data type from a string data type to an integer data type.
>>> **cups_consumed = int(input("How many cups did you drink? "))**
How many cups did you drink? **3**
>>> **type (cups _consumed)**
<class 'int'>
>>>

Python math operators

Python can be used to perform some basic calculations such as addition, subtraction, multiplication and division. You can use the python as a calculator.

To get an understanding of how python handles digits, simply open your python IDLE window and carry out some basic calculations from the command line. See below;
>>>>3 + 4
7
>>> 5 + 1
6
>>> 2 * 4
8

>>>16/2
8

```
File  Edit  Shell  Debug  Options  Window  Help
Python 3.8.5 (tags/v3.8.5:580fbb0, Jul 20 2020, 15:43:08) [MSC v.1926 32 bit (In
tel)] on win32
Type "help", "copyright", "credits" or "license()" for more information.
>>>
=== RESTART: C:/Users/user/AppData/Local/Programs/Python/Python38-32/math.py ===
>>> 3 + 4
7
>>> |
```

For division, python will automatically convert the result into a floating data type even though what you are dividing is an integer data type;

```
=== RESTART: C:/Users/user/AppData/Local/Programs/Python/Python38-32/math.py ===
>>> 3 + 4
7
>>>
=== RESTART: C:/Users/user/AppData/Local/Programs/Python/Python38-32/math.py ===
>>> 16/2
8.0
>>> |
```

Order of Operations

In fact, the Python programming language follows all the basic rules of mathematical calculations, including the order of operations. In the example below, Python performs the multiplication first follow by the addition operation:
>>> 3 + 5 * 5
28
>>>

And just like it is in mathematics, Python enables you to change the order of operations with the use of parentheses:

>>> (3 + 5) * 5
40
>>>

You can also nest parentheses as much as you need to in your calculations. Just ensure that you match up all the opening and closing parentheses properly. If you don't do this, python will not run as it will continue to wait for the parentheses that are missing. Check below;

>>> ((3 + 5) * 5

When you hit the Enter key, the IDLE sends back a blank line instead of giving you the result. It is waiting for you to close out the missing parenthesis. To run the command, just input the missing parenthesis on the blank line:

>>> (3 + 5) * 5)
40
>>>

CHAPTER TWO

MORE ABOUT PYTHON LANGUAGE

Data structures

The Python programming language supports four (4) essential data structures, which include list, tuple, set, and dictionary. These data structures feature a number of important built-in methods.

Lists

When you have values of a single or multiple data type, you can use lists to group these values together. The list data structure
can be assigned by specifying values using a square bracket together with a comma (,) as a separator:

>>> myList = ['a', 17, 'b', 12.5, 5, 2.9]
>>>print (myList)
['a', 17, 'b', 12.5, 5, 2.9]

```
=== RESTART: C:/Users/user/AppData/Local/Progr
['a', 17, 'b', 12.5, 5, 2.9]
>>> |
```

Just like strings, the values in a list can be checked using index numbers, which begins from 0. Python uses a feature named **slicing** to get a specified subset or element of the data structure by deploying the colon operator. You can specify slicing by using myList[start:end:increment] prompt. See the example below to better comprehend how slicing can be achieved;

-   You can get a single element in the list just like it is shown below:

    >>> myList[0]
    'a'

- You can get all the elements in a list by using empty to start and end values:

>>> myList[:]
['a', 17, 'b', 12.5, 5, 2.9]

```
=== RESTART: C:/Users/user/AppData/Local/Programs/Python/Python38-32/nhyy.py ==
>>> myList = ['a', 17, 'b', 12.5, 5, 2.9]
>>> myList[:]
['a', 17, 'b', 12.5, 5, 2.9]
>>>
```

- You can specify the start and the end index values to access a specific subset of the list:

>>> myList[1:5]
[17, 'b', 12.5, 5]

```
=== RESTART: C:/Users/user/AppData/Local/Programs/Python/Python38-32/nhyy
>>> myList = ['a', 17, 'b', 12.5, 5, 2.9]
>>> myList[:]
['a', 17, 'b', 12.5, 5, 2.9]
>>>


=== RESTART: C:/Users/user/AppData/Local/Programs/Python/Python38-32/nhyy
>>>
>>> myList[1:5]
[17, 'b', 12.5, 5]
>>>
```

- When you use the minus symbol together with an index number, you are telling the python interpreter to use that particular index number backwards. In the following instance, -1 backwards represents the index number 5:

>>> myList[1:-1]
[17, 'b', 12.5, 5]

```
=== RESTART: C:/Users/user/AppData/Local/Programs/Python/Python38-3
>>> myList = ['a', 17, 'b', 12.5, 5, 2.9]
>>> myList[:]
['a', 17, 'b', 12.5, 5, 2.9]
>>>

=== RESTART: C:/Users/user/AppData/Local/Programs/Python/Python38-3
>>>
>>> myList[1:5]
[17, 'b', 12.5, 5]
>>> myList[1:-1]
[17, 'b', 12.5, 5]
>>>
```

- You can get every other element in the list by
  providing the increment value with the start and the
  end values:

  >>> myList[0:5:2]
  ['a', 'b', 5]

```
=== RESTART: C:/Users/user/AppData/Local/Programs/Python/Python38-32/nk
>>> myList = ['a', 17, 'b', 12.5, 5, 2.9]
>>> myList[:]
['a', 17, 'b', 12.5, 5, 2.9]
>>>

=== RESTART: C:/Users/user/AppData/Local/Programs/Python/Python38-32/nk
>>>
>>> myList[1:5]
[17, 'b', 12.5, 5]
>>> myList[1:-1]
[17, 'b', 12.5, 5]
>>> myList[0:5:2]
['a', 'b', 5]
>>>
```

- You can know the length of a list variable by using the
  len() method. This method will be used often in most
  of the projects we will be doing in this guide.

  >>> len(myList)
  6

```
=== RESTART: C:/Users/user/AppData/Local/Programs/Python/Python38-32/nhyy
>>> myList = ['a', 17, 'b', 12.5, 5, 2.9]
>>> myList[:]
['a', 17, 'b', 12.5, 5, 2.9]
>>>

=== RESTART: C:/Users/user/AppData/Local/Programs/Python/Python38-32/nhyy
>>>
>>> myList[1:5]
[17, 'b', 12.5, 5]
>>> myList[1:-1]
[17, 'b', 12.5, 5]
>>> myList[0:5:2]
['a', 'b', 5]
>>> len(myList)
6
>>>
```

- You can add or delete an element from a list by using a number of operations. For instance, if you wish to add an element at the end of the list, utilize the append() method on the list:

>>> myList.append(10)
>>> myList
['a', 17, 'b', 12.5, 5, 2.9, 10]

```
=== RESTART: C:/Users/user/AppData/Local/Programs/Python/Python38-32/nhyy.py ===
>>>
>>> myList[1:5]
[17, 'b', 12.5, 5]
>>> myList[1:-1]
[17, 'b', 12.5, 5]
>>> myList[0:5:2]
['a', 'b', 5]
>>> len(myList)
6
>>> myList.append(10)
>>> myList
['a', 17, 'b', 12.5, 5, 2.9, 10]
>>>
```

You can see that an extra element 10 has been added to the list with the use of an "append" method.

- You can also add a particular element at a specific location by using the insert(i, x) method, where i stands for the index value, while x is the exact value that you want to add to the list:

>>> myList.insert(5,'hello')

>>> myList
['a', 17, 'b', 12.5, 5, 'hello', 2.9, 10]

```
=== RESTART: C:/Users/user/AppData/Local/Programs/Python/Pythor
>>>
>>> myList[1:5]
[17, 'b', 12.5, 5]
>>> myList[1:-1]
[17, 'b', 12.5, 5]
>>> myList[0:5:2]
['a', 'b', 5]
>>> len(myList)
6
>>> myList.append(10)
>>> myList
['a', 17, 'b', 12.5, 5, 2.9, 10]
>>> myList.insert(5,'hello')
>>> myList
['a', 17, 'b', 12.5, 5, 'hello', 2.9, 10]
>>>
```

In the script above, the "hello" has been added right after the number "5" because that is exactly what the script specified.

- In the same vein, you can utilize pop() to remove any element from the list. When you prompt only the pop() function without signifying the exact element that you want to remove, the pop function will remove the last element in the list. You will be able to remove a specific element from the list by using the pop(i) where 'i' is the particular element that you wish to remove.

>>> myList.pop()
10

```
=== RESTART: C:/Users/user/AppData/Local/Programs/Python
>>>
>>> myList[1:5]
[17, 'b', 12.5, 5]
>>> myList[1:-1]
[17, 'b', 12.5, 5]
>>> myList[0:5:2]
['a', 'b', 5]
>>> len(myList)
6
>>> myList.append(10)
>>> myList
['a', 17, 'b', 12.5, 5, 2.9, 10]
>>> myList.insert(5,'hello')
>>> myList
['a', 17, 'b', 12.5, 5, 'hello', 2.9, 10]
>>> myList.pop()
10
>>>
```

```
        >>> myList
    ['a', 17, 'b', 12.5, 5, 'hello', 2.9]
>>> myList.pop(5)
'hello'
>>> myList
['a', 2, 'b', 12.0, 5, 2]
=== RESTART: C:/Users/user/AppData/Local/Programs/P
>>>
>>> myList[1:5]
[17, 'b', 12.5, 5]
>>> myList[1:-1]
[17, 'b', 12.5, 5]
>>> myList[0:5:2]
['a', 'b', 5]
>>> len(myList)
6
>>> myList.append(10)
>>> myList
['a', 17, 'b', 12.5, 5, 2.9, 10]
>>> myList.insert(5,'hello')
>>> myList
['a', 17, 'b', 12.5, 5, 'hello', 2.9, 10]
>>> myList.pop()
10
>>> myList
['a', 17, 'b', 12.5, 5, 'hello', 2.9]
>>> myList.pop(5)
'hello'
>>> myList
['a', 17, 'b', 12.5, 5, 2.9]
>>>
```

Creating a List

There are about four (4) different methods of creating a list variable which include;

- Create an empty with an empty pair of square brackets. See below

```
>>> list1 = []
>>> print(list1)
```

[]
>>>

- Put a square bracket around a list of values that have
  been separated by comma. See below;
      >>> list2 = [1, 2, 3, 4]
      >>> print(list2)
          [1, 2, 3, 4]
      >>>

- Use the list() function to make a list from another
iterable object, see below:
      >>> tuple11 = 1, 2, 3, 4
      >>> list3 = list(tuple11)
      >>> print(list3)
          [1, 2, 3, 4]
>>>

-       Use a list comprehension.

The list comprehension system of creating lists is a bit more complicated way of creating a list from other data. Notice that with lists, Python deploys square brackets around the data values, not parentheses as with tuples (more on tuples later)

Just as with tuples, lists can feature any type of data, not just numbers or alphabets, as in this example:

>>> list4 = ['Rick', 'James', 'Katie Morgan', 'Jessica']
>>> print(list4)
['Rick', 'James', 'Katie Morgan', 'Jessica']
>>>

Extracting Data from a List

You can obtain individual data elements from list values by deploying index values. Check below using examples
>>> print(list2[0])
1
>>> print(list2[3])
4
>>>

You can as well use a negative number for specifying the list index. A negative index gets values starting from the end of the list:
>>> print(list2[-1])
4
>>>

You can observe that while using a negative index value, the -1 value begins from the end of the list because -0 is still 0. Notice that when you use negative index values, the -1 value starts at the end of the list, since -0 is
the same as 0.

Lists also agrees with the slicing method of getting a subset of the data elements which are contained in the list value, as in the following example:
>>> list4 = list2[0:3]
>>> print(list4)
[1, 2, 3]
>>>
The command above tells the print to display the list4 from 0 to 3 i.e from 1 to 2 to 3.

Tuples

Tuples, unlike lists, are immutable data structures that are supported by Python. Tuples being an immutable data structure implies that you cannot remove or add elements from or to the tuple data structure. Due to this immutable feature, it is faster and easier to access tuples unlike the mutable lists. Tuples are most often used to store sets of values that can never change. You declare a tuple data structure just the same way you declare lists but with the use of parentheses or without even using brackets at all. See example below;
>>> **tupleA = 1, 2, 3**
>>> **tupleA**
(1, 2, 3)

```
File   Edit   Shell   Debug   Options   Window   Help
Python 3.8.5 (tags/v3.8.5:580fbb0, Jul 20 2020, 15:
tel)] on win32
Type "help", "copyright", "credits" or "license()"
>>>
== RESTART: C:/Users/user/AppData/Local/Programs/Py
>>> tupleA = 1, 2, 3
>>> tupleA
(1, 2, 3)
>>>
```

>>> **tupleB = (1, 'a', 3)**
>>> **tupleB**

(1, 'a', 3)

```
File   Edit   Shell   Debug   Options   Window   Help
Python 3.8.5 (tags/v3.8.5:580fbb0, Jul 20 2020,
tel)] on win32
Type "help", "copyright", "credits" or "license
>>>
== RESTART: C:/Users/user/AppData/Local/Program
>>> tupleA = 1, 2, 3
>>> tupleA
(1, 2, 3)
>>> tupleB = (1, 'a', 3)
>>> tupleB
(1, 'a', 3)
>>>
```

Just like you have it in a list data structure, the values in tuple can be called using index numbers:

>>> tupleB[1]

'a'

```
== RESTART: C:/Users/user/AppData/Local/Pro
>>> tupleA = 1, 2, 3
>>> tupleA
(1, 2, 3)
>>> tupleB = (1, 'a', 3)
>>> tupleB
(1, 'a', 3)
>>> tupleB[1]
'a'
>>>
```

Since tuples are not exactly mutable like lists, most of the manipulations you carried out for lists (such as append, insert and pop) cannot be done for lists.

Sets

The set data structure in Python is designed to aid mathematical set operations. The set data structure features an unordered assemblage of elements without duplicates. When you have a list of numbers and some of the numbers appear more than once, you can use the set() function to remove the ones that are duplicate and then have each of them appearing once.

For instance

```
>>> listA = [1, 2, 3, 1, 4, 5, 3, 2]
>>> setA = set(listA)
>>> setA
set([1, 2, 3, 4, 5])
```

Dictionaries

The dict data structure is deployed to store key-value pairs indexed by keys, which are also identified in other languages as associative arrays, hashes, or hashmaps. Unlike other data structures, you can extract dict values by using associated keys:

```
>>> boards = {'uno':328,'mega':2560,'lily':'128'}
>>> boards['lily']
'128'
>>> boards.keys()
['lily', 'mega', 'uno']
```

Controlling the flow of your program

Just like any other language, Python enables you to control the program flow using some compound statements. In this section, we will briefly introduce these statements to you.

The if statement

The *if statement* is perhaps the most simple and standard statement you can use to set up conditional flow. To have a better understanding of the if statement, prompt the following code in the Python interpreter while using different values of the age variable:

```
>>> age = 15
>>> if age < 17 and age > 12:
print "Teen"
elif age < 13:
print "Child"
else:
print "Adult"
```

This will print teen as your result on the screen.

The for statement

The Python's for statement iterates over the elements of any sequence based on the order of the elements in that sequence:

```
>>> celsius = [13, 21, 23, 8]
>>> for c in celsius:
print " Fahrenheit: "+ str((c * 1.8) + 32)
```

The python interpreter will produce the following result that will show the calculated values in Fahrenheit from the provided Celsius values.

```
Fahrenheit: 55.4
Fahrenheit: 69.8
Fahrenheit: 73.4
Fahrenheit: 46.4
```

The while statement

You can utilize the While statement to produce a continuous loop in a python language. A while loop will continue to iterate over the code block until the condition is proved true:

```
>>> count = 5
>>> while (count > 0):
print count
count = count - 1
```

The while statement will continue to iterate and print the value of the count variable and also reduce its value by 1 until the condition, that is (count > 0), is true. As soon as the value of count is smaller than or equal to 0, the while loop will exit the code block and stop the iteration process.

Built-in functions

Python supports many useful built-in functions that do not need the user to import any external libraries.

Conversions

The conversion parameters such as int(), float(), and str() can convert other categories of data into integer, float, or string data types respectively:

```
>>> a = 'a'
>>> int(a,base=16)
10
```

```
>>> i = 1
>>> str(i)
'1'
>>> a = 'a'
>>> int(a,base=16)
10
>>>
```

```
>>> i = 1
>>> str(i)
'1'
```

You can also use parameters such as list(), set(), and tuple() to convert one data structure into another data structure.

Math operations

Python features a built-in mathematical function that can process the minimum and/or maximum values from a list. See below;
>>> list = [1.12, 2, 2.34, 4.78]
>>> min(list)
1.12
```
>>> list = [1.12, 2, 2.34, 4.78]
>>> min(list)
1.12
>>>
```
>>> max(list)
4.78
```
>>> list = [1.12, 2, 2.34, 4.78]
>>> min(list)
1.12
>>> max(list)
4.78
>>>
```

You can also carry out other mathematical calculations such as addition, subtraction, multiplication and division
```
list = [1.12, 2, 2.34, 4.78]
>>> min(list)
1.12
>>> max(list)
4.78
>>> max(list) / min(list)
4.267857142857142
>>>
Max(list) * min(list)
5.3536
>>>
Max(list) – min(list)
3.66
```

```
>>>
Max(list) + min(list)
5.9
>>>
```

```
>>> list = [1.12, 2, 2.34, 4.78]
>>> min(list)
1.12
>>> max(list)
4.78
>>> max(list) / min(list)
4.267857142857142
>>>
>>> max(list) * min(list)
5.353600000000001
>>> max(list) - min(list)
3.66
>>> max(list) + min(list)
5.9
>>> |
```

The pow(x,y) function solves the value of x raised to the power of y. for instance, pow(max(list), min(list)) above will return 5.767123400417216

Also,

>>> pow(3, 2)

9

```
>>> pow(max(list), min(list))
5.767123400417216
>>> pow(3, 2)
9
>>> |
```

String operations

Python gives easy access to string manipulation through the built-in functions that have been fully optimized for performance. See below;

>>> str = "Hello World!"

>>> str.replace("World", "Universe")

'Hello Universe!'
See below code to split a string that has a separating character where the default character is space:
>>> str = "Hello World!"
>>> str.split()
['Hello', 'World!']
See the code below to split a string from a separating character for any other character:
>>> str2 = "John, Merry, Tom"
>>> str2.split(",")
['John', ' Merry', ' Tom']
See the code below to convert an entire string value into uppercase or lowercase:
>>> str = "Hello World!"
>>> str.upper()
'HELLO WORLD!'
>>> str.lower()
'hello world!'

You can deploy the built-in open function to access a file in a python script. The basic syntax for this is written below;
filename_variable = open (*filename, options*)

Closing a File

When you open a file in any program, it is only safe you close the file before you exit the program. The same scenario applies in a python script. To close an opened file in python, use the syntax below;
filename_variable.close *()*

Writing to a File

You can either open a file for the purpose of writing alone, or you can open the said file to be read and written. The open mode usually deployed for writing a text file is either w or a, which actually depends on whether you want to create a new file or append to an already existing file. Adding a + sign at the end of the w or a open mode will enable you to both read and write to the file.

Writing to a Pre-existing File

You instruct Python that your written data will be appended to a pre-existing file through the open function. After that, you can now write data to a pre-existing file by using the .write just like the way you used the .write method to write data to a new file.

CHAPTER THREE

INTRODUCTION TO ARDUINO

Arduino is not just a name of a device or a computer as you might have perceived. More than a gadget, Arduino is the software and Hardware Company that makes single-board microcontrollers that can be used for manufacturing digital devices. It is an open source company comprising builders, users and project communities that come together for the purpose of designing a single-board computer. One example of such a computer is the Arduino Uno. The Arduino board and its software are licensed by the manufacturer under the CC-BY-SA license and the GNU General Public License. These licenses give permission to any company or individual, which is not the initial maker of Arduino, to manufacture both the software and hardware (Arduino board) without penalty. If you are a builder or an engineer, and you are currently reading this book, you can actually design your own board just like an Arduino board; no one will penalize you. The Arduino board comprises some set of digital and analog input/output (I/O) pins that can be used for interfacing with other boards and circuits. The board is also made up of serial communication interfaces such as the Universal Serial bus (USB) which can be used to connect Arduino with other computers. The Arduino project was first initiated in 2005 and was meant as a tool to enable students and novices to design devices and materials that will be able to interact, at the basic level, with the environment. Thermostats, simple robots and motion detectors are examples of devices that can interface with the Arduino project. The Arduino project started as a team work with Massimo Banzi, David Cuartielles, Tom Igoe, Gianluca Martino, and David Mellis.

Most Arduino boards are made up of an Atmel 8-bit AVR microcontroller (ATmega8, ATmega168, ATmega328, ATmega1280, or ATmega2560) featuring different levels of flash memory, features and pins. The 32-bit Arduino Due, based on the Atmel ARM-based processor (SAM3X8E) came into the market in 2012. The boards deploy single or double-row pins or a female header, which can facilitate connections for programming and attachment into other circuits. The single or double-row pins or a female header can connect with add-on modules called shields. Most boards feature a 5 V linear regulator with a 16 MHz crystal oscillator or ceramic resonator. Some designs, like the LilyPad, operate at 8 MHz and dispense with the onboard voltage regulator. Arduino microcontrollers are already pre-programmed with a boot loader that helps simplify uploading of various programs to the on-chip flash memory. The default bootloader that came with the Arduino Uno is called the Optiboot bootloader. Boards are installed with program code through a serial connection to work with another computer. Current Arduino boards are organized via Universal Serial Bus (USB), perfected using USB-to-serial adapter chips like the FTDI FT232. Some boards, such as the later-model Uno boards, replace the FTDI chip with a separate AVR chip featuring USB-to-serial firmware, which is reprogrammable via its own in-system programming header. Other Arduino variants, like the Arduino Mini and the unofficial Boarduino, deploy a detachable USB-to-serial adapter board or cable, Bluetooth or other methods. When the other variants are used with traditional microcontroller tools, instead of the Arduino IDE, the standard AVR in-system programming (ISP) programming is utilized.

The Arduino Uno: the most popular Arduino model

The Arduino Uno is the most popularly used of the Arduino series. If you are a beginner who is just getting started with coding and electronics, this is the board for you. **Arduino Uno** is essentially a microcontroller single-board designed on the ATmega328P platform. It features 14 digital input/output pins out of which six (6) can be deployed for pulse-width modulation output (PWM), 6 analog inputs, a 16 MHz ceramic resonator (CSTCE16M0V53-R0), a Universal Serial Bus (USB) connection, a power jack, an in-circuit serial programming header (ICSP) and a reset button. The Universal Serial Bus (USB) provides a platform for connecting a circuit device or another computer. The in-circuit serial programming header (ICSP) gives the microcontroller its ability to be programmed without even disconnecting it from the circuitry. You can simply connect your Arduino to a computer using a Universal Serial Bus (USB) cable or power it using a battery or an adapter.

The Arduino Uno variants

Depending on the project that you want to do, hardware specifications depend largely on project requirements. If you are working on a project that needs you to interface with a number of external parts, you will require a prototyping platform that features a sufficient number of **input/output (I/O)** pins that you can use for interfacing purposes. In the same vein, if you are working on a project that requires you to carry out a large array of complex mathematics, you will need a platform with large computational ability. That said, the Arduino board can be found in 16 different official models, and each model of the Arduino board differs from the others when it comes to form factor, input/output pins, computational capability and many other on-board specifi-cations. The Arduino Uno is the simplest and most popular model, which is good enough for simple Do-It-Yourself projects. We will be using the Arduino Uno board for most of the projects in this guide. The Arduino Mega board is another widely used variant, and it is very big with more I/O pins and a very powerful microcontroller. Find below some of the major Arduino variants with their specifications;

1. **Arduino Uno**
   - **Processor: ATmega328**
   - **Processor frequency: 16 MHz**
   - **Digital Input or output:14**
   - **Digital Input or output with PWM: 6**
   - **Analog Input or output: 6**
2. **Arduino Leonardo**
   - **Processor: ATmega32u4**
   - **Processor frequency: 16 MHz**
   - **Digital Input or output:14**
   - **Digital Input or output with PWM: 6**
   - **Analog Input or output: 12**
3. **Arduino mega**
   - **Processor: ATmega2560**
   - **Processor frequency: 16 MHz**
   - **Digital Input or output:54**
   - **Digital Input or output with PWM: 14**
   - **Analog Input or output: 16**
4. **Arduino Nano**
   - **Processor: ATmega328**
   - **Processor frequency: 16 MHz**
   - **Digital Input or output:14**
   - **Digital Input or output with PWM: 6**
   - **Analog Input or output: 8**

5. **Arduino Due**
   - **Processor: AT91SAM3X8E**
   - **Processor frequency: 84 MHz**
   - **Digital Input or output:54**
   - **Digital Input or output with PWM: 12**
   - **Analog Input or output: 12**
6. **LillyPad Arduino**
   - **Processor: ATmega168v**
   - **Processor frequency: 8 MHz**
   - **Digital Input or output :14**
   - **Digital Input or output with PWM: 6**
   - **Analog Input or output: 6**

You can program any one of the variants discussed above by deploying a common integrated development environment, which you can refer to as **Arduino IDE.** The type of project that you want to do will inform your choice of the Arduino board as you can actually get to choose between any of the variants. Your Arduino integrated development environment (IDE) should be able to write and download any Arduino program you select on to the board.

Installing the Arduino Integrated Development Environment (IDE)

Just like installing, the Python IDE is the first thing to get started with the python, installing the Arduino integrated development environment (IDE) on your computer is the first step to getting started with Arduino. You can remember when you want to install the python IDE; you installed the IDE based on the operating system you are running on your computer. The same thing will apply with the installation of the Arduino IDE. You will install the Arduino based on your operating system. See the guides below;

Linux OS

You don't really need too much stress to install the Arduino integrated development environment in your Ubuntu operating system. This is because the repository of your Ubuntu OS already has the Arduino IDE with the needed dependencies. If you are using the Ubuntu 12.04 or the newer version of the Ubuntu, you can install the Arduino from the terminal by prompting the command below;

$ sudo apt-get update && sudo apt-get install arduino arduino-core

The Ubuntu repository has the latest version of the Arduino integrated development environment (IDE) as version 1.8.12. If you are using the Fedora 17 or the latest version of RedHat Linux, you can prompt and execute the following script in the terminal:

$ sudo yum install arduino

You can also install the Arduino IDE on Ubuntu with the DEB package. The Arduino DEB packages are not readily available in the Ubuntu default repositories. If you want to install it, you need to download the .DEB package from its download's page at https://www.arduino.cc/en/Main/Software. You can install the .DEB package on both the 32-bit version and the 64-bit version of Ubuntu. Alternatively, you can deploy the following wget command to get the Arduino Software (IDE) package directly on the terminal. The current latest version of the Arduino is 1.8.12.

cd /tmp
wget https://downloads.arduino.cc/arduino-1.8.12-
linux64.tar.xz

The next step will be to extract the downloaded archive file by deploying the tar command as displayed below.

tar -xvf arduino-1.8.12-linux64.tar.xz

You can then move into the **Arduino-1.8.12** which has been extracted and run the installation script as root.

cd arduino-1.8.12/
sudo ./install.sh
Once the installation has been completed, a desktop icon will then appear on your desktop.

You can also install the Arduino IDE by using Snap. Snaps are software packages that you can easily create and install to manage and use packages on the Linux distributions system. The Snap provides the easiest and most organized way of installing and managing packages on the Linux distributions. You can prompt the command below to install Snap packages and then install the Arduino IDE;

sudo apt install snapd
sudo snap install arduino
In order to allow code uploading on the Arduino board over USB, you will be required to add your user to the dialout group and then connect the snap to the USB socket (raw-USB socket). Simply open the terminal, run the prompt below and then restart your computer.

sudo usermod -a -G dialout $USER
sudo snap connect arduino:raw-usb
You can then restart the computer to get started.

Mac OS X

If you want to install the Arduino integrated development environment on Mac OS X (10.7 or newer), follow the guides below;

1. Get the updated version of the Arduino integrated development environment (IDE) for Mac OS X from http://arduino.cc/en/Main/Software.
2. Unzip the downloaded Arduino IDE and then drag it to the application folder.
The Arduino IDE for Mac OS is incorporated with Java and you will be required to get the latest version of Java on your computer. If you don't have the upgraded version of Java on your device, the program (Arduino IDE) will ask you upon installation to install the Java SE 6 runtime. Follow the request and install Java as the Mac OS will help you to install it automatically.

Windows

Installation of Arduino for your Windows is not really difficult. All that you need to do is to download the setup file from http://arduino.cc/en/Main/Software. Choose the most recent version of the Arduino integrated development environment (IDE), that is, 1.0.x or a newer version.

Verify the operating system on your Windows, and then download the right Arduino model based on your operating system, which can either be 32 bits or 64 bits. Install the Arduino integrated development environment (IDE) to the default location as already indicated in the installation wizard. Once you have successfully installed the Arduino IDE, you can navigate to **start| programs** on your computer to get started.

Getting started with the Arduino IDE

The Arduino IDE is a cross-platform application created in Java that can be deployed to build, compile, and upload programs to your Arduino board. When you launch your Arduino IDE, you will see an interface that is similar to the interface shown in the following screenshot.

The integrated development environment (IDE) features a text editor which can be used for coding purposes, a menu bar to have access to the IDE components, a toolbar to easily have access to the most common IDE functions, and a text console to see the compiler outputs. Check the status bar at the bottom to show the selected Arduino board and the name of the port that it is connected to. The screenshot below show the Arduino interface;

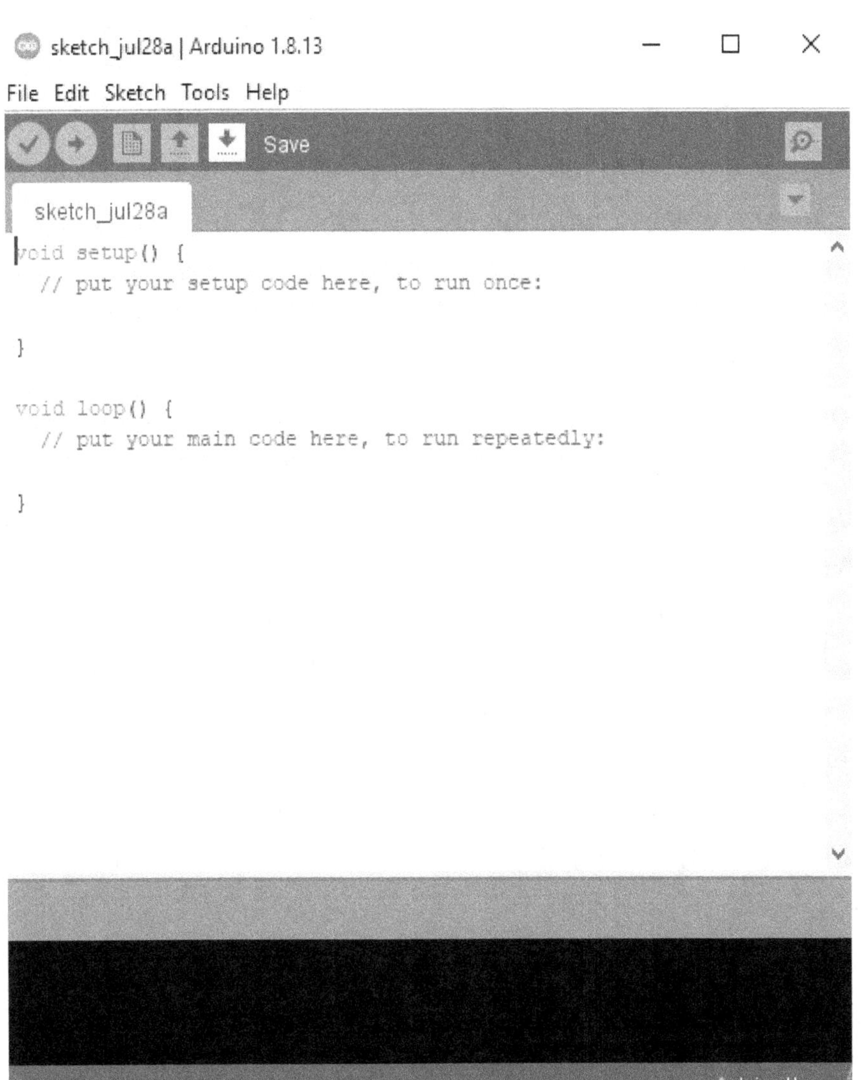

The bottom right hand side of the screenshot above shows the type of Arduino board which the software is currently connected to. Likewise, the top left section is showing the sketch name together with the date the software is being used. Today being July 28, 2020, this is how your own software will be showing the current date anytime you open it.

Arduino Sketch

The Arduino sketch is an Arduino program that has been developed solely for you using the Arduino integrated development environment (IDE). The Arduino sketches are written in Arduino language which is a custom model of the javascript. The Arduino sketch has a built-in text editor where you can write, edit and even copy & paste your codes. As soon as you finish writing the code, you can then save the code with the **.ino** extension. The **.ino** extension is the format which the latest version of the Arduino saves texts with. Anytime you save your code or texts file, the Arduino IDE will instantly create a folder for you to save the text files. The location where the file has been saved is called a **sketchbook.** Even if you are using some other supporting files for your sketch, like the library files or header file, they will still be saved in the **sketchbook** location. You can open a new sketchbook by choosing **new** from the **file** bar as shown below;

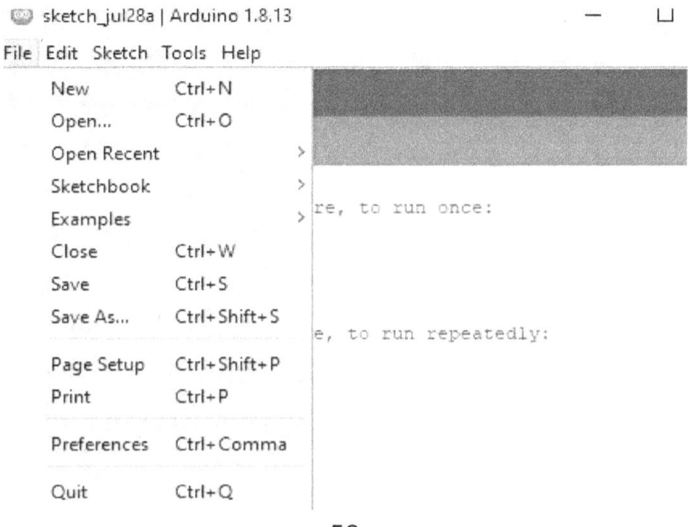

You can alternatively press **CTRL + N** on your computer to open a new sketchbook.

Working with the Arduino library

The Arduino IDE makes use of the libraries to gain access to more functionality from your existing sketches. The libraries are some set of functions that are able to carry out some tasks around a specified concept or component. You can start working with any external hardware component right from the software by using some set of customized Arduino libraries. You can import library from the **sketch** menu and then scroll down to select **Include Library.** See screenshot below;

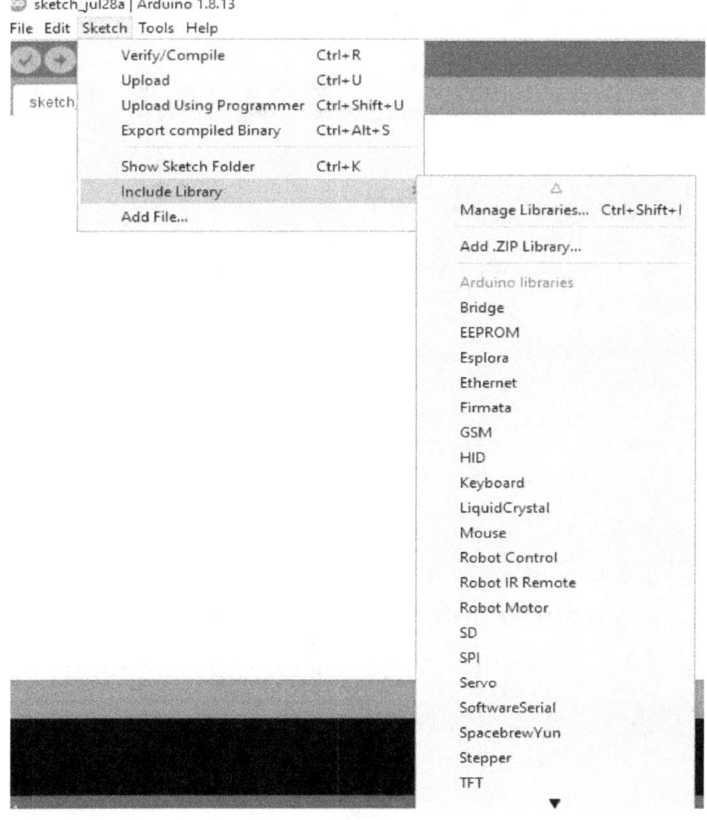

You can get to use any library out of the multitude of libraries in your sketch by merely specifying the library you want to use using the **#include statement** at the start of the sketch. For instance, you can include the bridge library by typing **#include <Bridge.h>.**

The Arduino Built-in example sketches

The Arduino IDE features a huge array of built-in example sketches. These examples are there to make you get familiar with some basic Arduino concepts and libraries. The Arduino community has a good support base for each of the examples you see under the Arduino example section. To have access to the examples from the Arduino IDE, simply tap on **Files** and then scroll down to select **Examples.** See below;

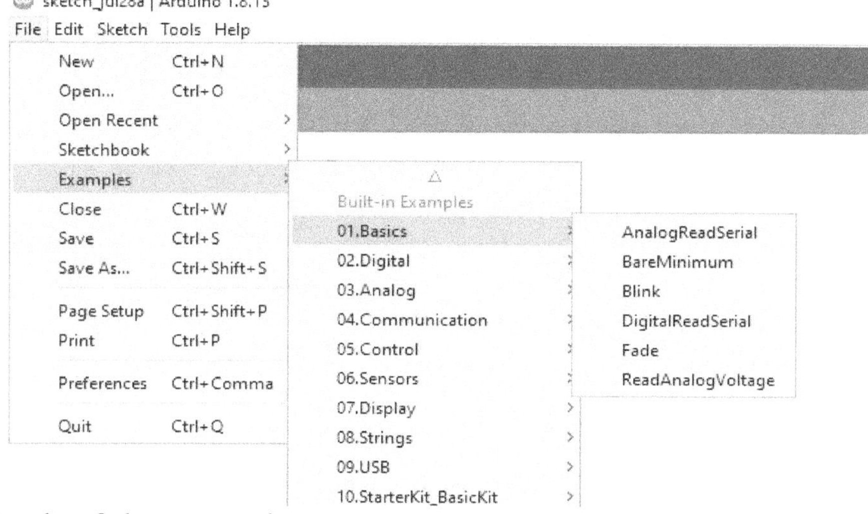

Each of the examples you are seeing in the screenshot above shows basic Arduino commands which can be used for a lot of things like turning on/off LED light, reading a poten-tiometer etc.

If you toggle down the **basic** example menu, you will see some examples like the one below;

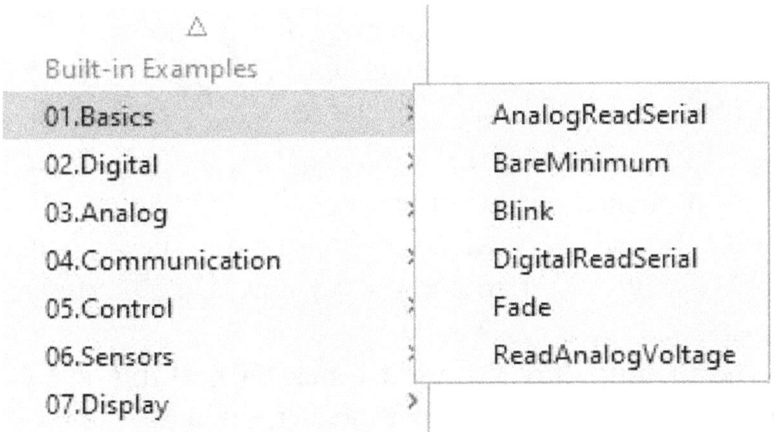

Each of these examples ranging from **AnalogReadSerial** to **ReadAnalogVoltage** can be used as follows;

- **Analog Read Serial:** You can use this to read a potentiometer, print the state of the potentiometer out to the Arduino Serial Monitor.
- **Bare Minimum:** This is the bare minimum of code you would need to start your Arduino sketch.
- **Blink:** You can use this to turn an LED on and off.
- **Digital Read Serial:** You can use this to read a switch, print the state of the switch out to your Arduino Serial Monitor.
- **Fade:** Shows you how to use the analog output to fade an LED.
- **Read Analog Voltage:** You can use this to read an analog input and then print the voltage result out to the Serial Monitor.

If you select the **Digital Example, you see the following;**

- **Blink Without Delay:** You can use this to blink an LED without necessarily using the delay() function.
- **Button:** Use the pushbutton to control the LED.
- **Debounce:** Read a pushbutton and filter the noise.
- **Digital Input Pullup:** shows you the use of INPUT_ PULLUP with pinMode().
- **State Change Detection:** You can use this to count the number of pushes of the button.
- **Tone Keyboard:** A musical keyboard using piezo speaker
- **Tone Melody:** You can use this to play a melody using a Piezo speaker.
- **Tone Multiple:** Play different tones on multiple speakers sequentially by deploying the tone() command.
- **Tone Pitch Follower:** You can use this to play a pitch on a piezo speaker depending essentially on the input from an analog.

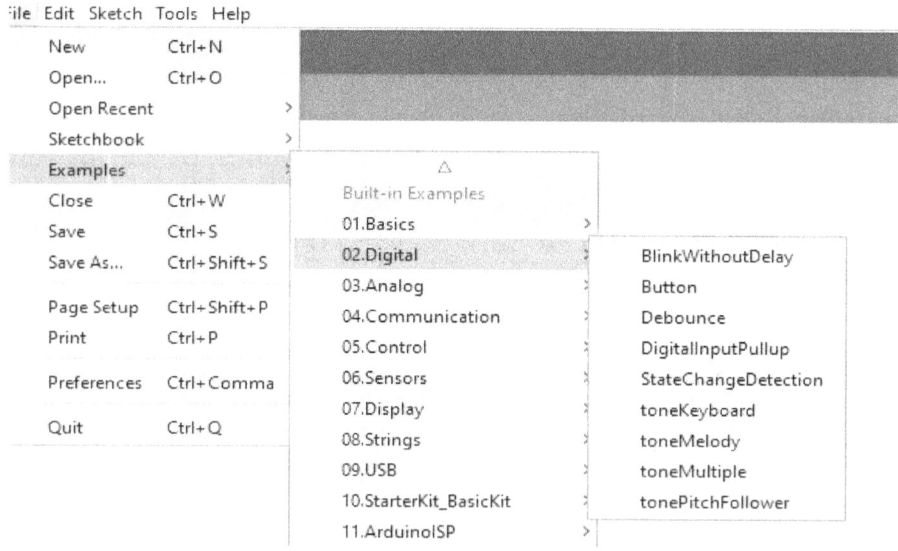

Let us see a built-in example of how to turn on/off LED.

You should start by building the circuit plug of your Arduino board into your computer, then initiate the Arduino Software (IDE) and prompt the code below. You may also load these codes right from the menu File by selecting Basics and then tapping on Blink.

The first thing you should try is to initialize LED_BUILTIN pin as an output pin with the line below

```
pinMode(LED_BUILTIN, OUTPUT);
```

Turn on the LED light in the main loop with the code below;

```
digitalWrite(LED_BUILTIN, HIGH);
```

When you prompt the code above, you will be supplying 5 volts to the LED anode. This will create a voltage difference across the pins of the LED, and then lights up your LED.

You can then turn it off with the code:

```
digitalWrite(LED_BUILTIN, LOW);
```

The code above will take the **LED_BUILTIN** pin back to 0 volts, and then turns the LED light off. In between the on and the off session, you want much time that will allow the person or user see what is happening (turning on/off), the delay command is the one that will Arduino board to do nothing for a very small seconds. Therefore, when you deploy the delay () command, nothing else will happen for that specific amount of period.

When you tap on the blink example from the basic option, the Arduino IDE will give you a new window containing the code below;

```
// the setup function runs once when you press reset or power the board

void setup() {

  // initialize digital pin LED_BUILTIN as an output.

  pinMode(LED_BUILTIN, OUTPUT);

}

// the loop function runs over and over again forever

void loop() {

  digitalWrite(LED_BUILTIN, HIGH);   // turn the LED on (HIGH is the voltage level)

  delay(1000);                       // wait for a second

  digitalWrite(LED_BUILTIN, LOW);    // turn the LED off by making the voltage LOW

  delay(1000);                       // wait for a second

}
```

CHAPTER FOUR

CHOOSING YOUR ARDUINO BOARD

Now that you have opened the code in the integrated development environment (IDE), you can on-board the type of board you want to be using. This board is where you are going to be uploading all of your sketches. The Arduino integrated development environment (IDE) will need to know your type of board. This is so that the board will be able to implement the program for the right microcontroller.

To choose your type of Arduino board, simply scroll to the **Tools** section and then select **board.** See below;

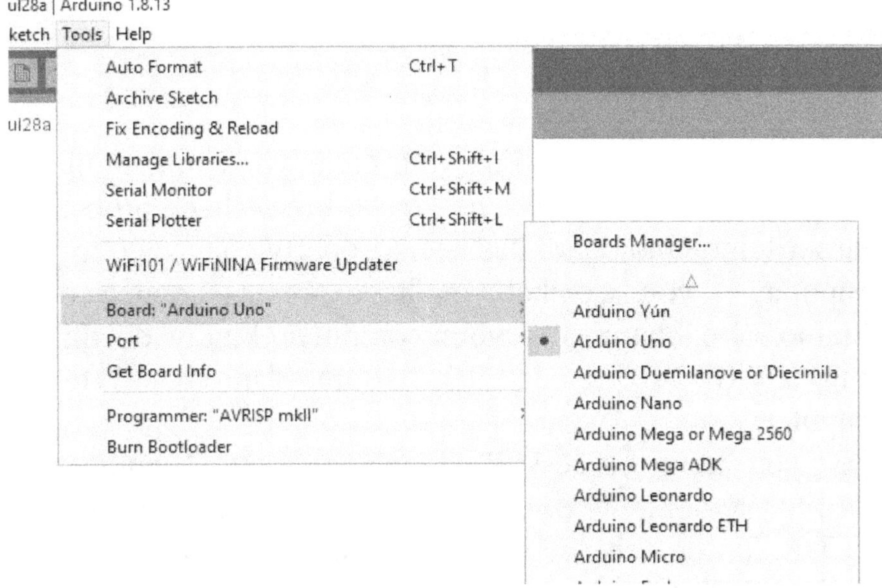

Unless you have another type of Arduino board that is not the Arduino Uno board, simply select the Arduino Uno board. After selecting the board, you can then start compiling your sketches.

You can start compiling your sketches by tapping on the **Sketch** menu and then select **Verify or compile** from the top menu bar. Alternatively, you can press **CTRL + R** on your computer to start compiling. This will enable you to start compiling your codes if there is no error in any of your connections.

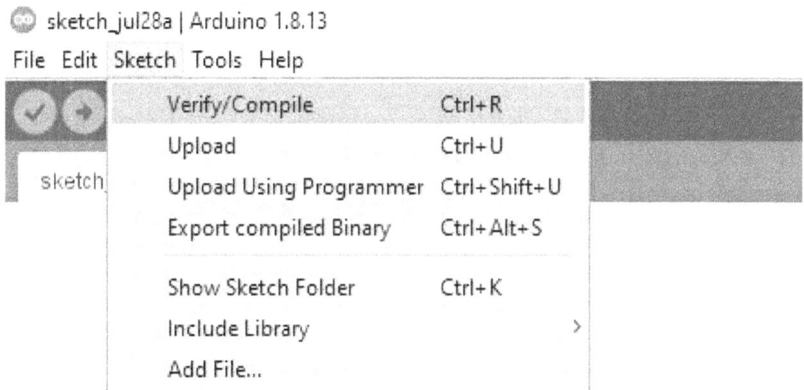

Once you are done compiling your code, you can then upload the code to your Arduino Uno board. First, check properly if your Arduino Uno board has been properly connected to your computer. If you see that you have not yet connected the Arduino Uno board with your computer, kindly do this by using a USB to connect the Arduino Uno board to the computer.

After you have successfully connected the Arduino Uno board to the computer, you can then select the serial port for the IDE. To do this, move your cursor to the **Tools** section and then **Serial Ports.** Select the appropriate serial port from the serial port menu.

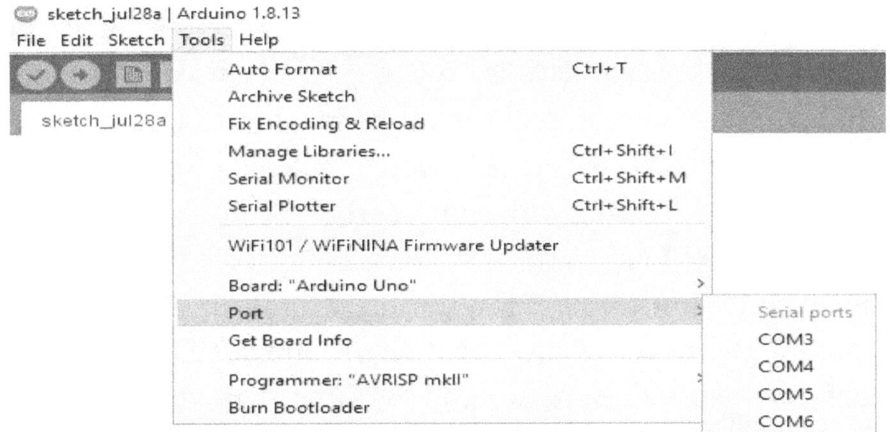

Note that if you are using some Linux distributions, it is possible that you will not be able to upload your Arduino program to the Arduino board. This is due to some permission restrictions on the serial port. You can bypass this restriction by prompting the command below on the terminal;

$ sudo usermod -a -G uucp, dialout, lock <username>

You can now start to upload your compiled sketch to your Arduino Uno board by tapping on the **File** menu and then select **Upload.**

This will deploy the serial connection to enter the compiled firmware in the Arduino microcontroller. Please exercise a little patience until you observe that the LED light on the board has stopped flashing. With this, your Arduino board is ready for sketches.

You can monitor or see the action of the blinking LED near digital pin 13.

Using the Serial Monitor window

When you wanted to board your Arduino Uno onto your computer, we used the Universal Serial Bus (USB) cable to attach the Arduino board into the computer. The Universal Serial Bus (USB) is the standard, which gives you the opportunity to attach many electronic parts to your computer by using the serial interface. When you attach your Arduino Uno board to your computer with the USB, the computer will notice the board as a peripheral gadget. This type of connection you made with your USB will be referred to as **Serial connection.**

The **Serial Monitor** window is just a built-in utility of the Arduino integrated development environment (IDE). To have access to the serial monitor window, simply move your cursor to **Tools** and then tap on **Serial Monitor.** You can alternatively press **CTRL + SHIFT + M** on your computer to have access to the serial monitor window. You can configure the serial monitor window to be able to monitor data or information that is being sent or received on the serial port that was used to connect the Uno board to your computer.

The baud rate of the serial connection can also be configured with the drop-down icon. This feature is especially handy when you want to test your prototypes for their performances.

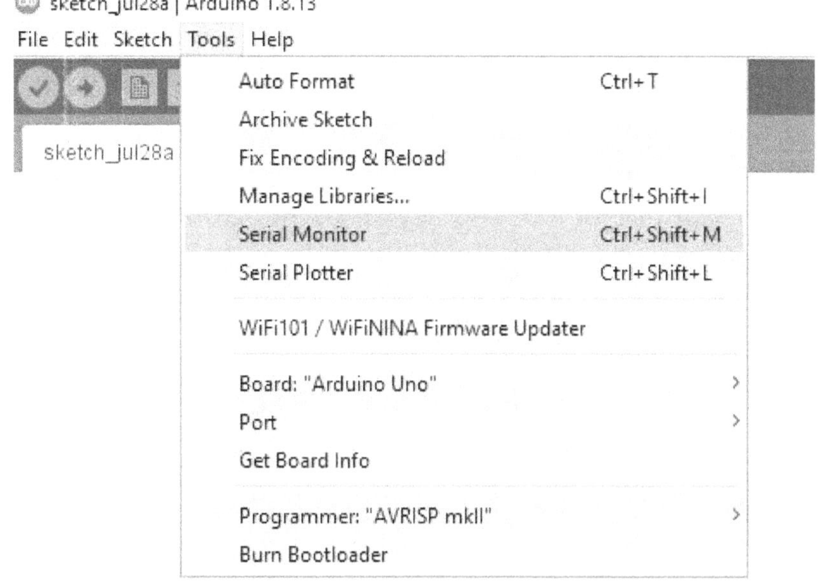

Uploading Your First Sketch

The Arduino equivalent of the "Hello World" which is default in most traditional programming languages is the blinking LED. Let us begin first by installing the blinking LED on the Arduino board followed by slight modification to make the light blinks faster. Your Arduino comes up with an empty sketch (containing no sketch) when you initially open the Arduino software. Navigate to the **File** menu and open the Blink example (this has been discussed in the previous concept). You will be uploading the Blink sketch onto your Arduino Uno board. Use the USB cable to first connect your Arduino board to the computer. After you have successfully plugged the devices, you should observe the green LED light on the Arduino coming up. The Arduino board will, by now, be flashing with light. This is because the boards are now pre-installed with the Blink sketch. You can install the blink sketch again so that you can be able to do some little modifications. Mac users, upon successfully plugging the board, may be prompted with a message "A new network message has been detected", just cancel this message as it is just a notification telling you that you have inserted a device into your computer. You won't be able to upload your sketch successfully until you let the Arduino software the type of Arduino board that you are using and the serial port you are connecting with. Kindly go to the **Tools** menu to select the serial port and the board type. If you are using a Windows computer, your serial port will always be COM3. Select the Upload icon in the toolbar, once you select the icon, you will notice a short pause while compiling the sketch after which sketch uploading will then begin. Once the uploading is completed, you will receive a "Done Uploading" message at the bottom side of the window.

Upon successful uploading, the Arduino board will start running the sketch automatically for you and you will observe that the LED light will start to blink. If you don't see any result, kindly check your serial and board type settings. You can modify this sketch to enable the LED light blink faster by simply modifying the two places in your sketch where there is a short pause of about 1000 milliseconds so that the short pause is now about 500 milliseconds. To do this, kindly modify the blink script as follow;

```
Void setup()
{
  // initialize digital pin LED_BUILTIN as an output.
  pinMode(LED_BUILTIN, OUTPUT);
}


// the loop function runs over and over again forever
void loop() {
  digitalWrite(LED_BUILTIN, HIGH);   // turn the LED on (HIGH is the voltage level)
  delay(1000);               // wait for a second
  digitalWrite(LED_BUILTIN, LOW);    // turn the LED off by making the voltage LOW
  delay(1000);               // wait for a second
}
```

Change the delay time above from 1000 to 500.

```
Void setup()

{

  // initialize digital pin LED_BUILTIN as an output.

  pinMode(LED_BUILTIN, OUTPUT);

}


// the loop function runs over and over again forever

void loop() {

  digitalWrite(LED_BUILTIN, HIGH);   // turn the LED on (HIGH is the voltage level)

  delay(500);                        // wait for a second

  digitalWrite(LED_BUILTIN, LOW);    // turn the LED off by making the voltage LOW

  delay(500);                        // wait for a second

}
```

PROGRAMMING THE ARDUINO

The C Language is the language of instruction for the Arduino board. This section will only teach you what you need to know to use the C language to be able to program your Arduino. You will be using the basics here in every Arduino sketch you make. Programming, unlike most languages we speak, is not innate in us as it is not our mother's tongue. To get the best out of any programming language, you need to dedicate your time and attention to it. If you have used a programming language, such as python or any other programming language, you will easily get the basics of the C programming language. The programs in Arduino are called sketches. You can see these sketches as instructions you need to carry out in the order in which they appear. For instance, let us say you are supposed to carry out the instruction below; Note that the pin 13 in the code below stands for **LED_BUILTIN (LED light on Arduino is on pin 13).**

digitalWrite(13, HIGH);

delay(500);

 digitalWrite(13, LOW);

The three lines you are seeing above are different lines and they each carry out a task. The first line is an instruction to set the output on pin 13, which is the LED light, to high mode. So the LED light will literally come up at this stage. The second line is an instruction to wait for 500 milliseconds, while the third line will turn off the LED light. These three lines, consecutively, will carry out the task of getting the LED light to blink a single time.

Let us begin by getting the basics. The fear and frustration of most beginners is not that they don't know what to write when it comes to syntax; their fear is in how to write it properly. The first thing you should take care of as a beginner in syntax is proper punctuation and the ways the words in codes are weaved. You write the name of a single thing as a single word (if your intention is to write the name of one thing). This implies that you are not expected to put a space between the letters. In the example code above, the **digitalWrite** is a name of a specific built-in function whose job is to set an output pin (pin 13) on your Arduino board.

 While writing names in syntax, you must understand that the programming language is case sensitive. digitalWrite and

DigitalWrite or digitalwrite are three different things entirely. While digitalWrite will carry out the task specified above, DigitalWrite will not. The digitalWrite function needs to be informed which pin on the Arduino you want to set and whether to take it high or to take it low. These two information are referred to as arguments and the program will help you pass these arguments to a function when called. You should separate the parameters in the function with a parenthesis and set off by a comma. The normal is to put the first parentheses (opening parentheses) just immediately after the last letter in the function's name and then to insert a space right after the comma before the following parameter. However, you can insert space characters inside the parentheses if you desire. If there is only one argument in the function, then you don't need to put a comma. You can observe how a semicolon ends each line, but it would be preferable if it were a period. This is because the semicolon marks the finishing of a statement. In the following section, you will be able to understand more about what happens when you hit the Upload button on your Arduino IDE. This will enable you to try a few do-it-yourself projects.

Getting a grip of how codes perform real action on Arduino

We can now recognize an Arduino sketch and of course have a rough hint of what the Arduino sketch is trying to do, but we still have to see deeper into how most of these programming language codes we are talking about transform from being words on some pages to instructions that carry out something real, such as turn an LED on and off. The processes involved in compiling your codes from the Arduino software text editor and deploying them in real scenarios on the Arduino board is an interesting one. To get started, the computer carried out a step called **compilation** where it exports the codes you have written and then translates it into a machine language which the Arduino understands (the machine languages are some set of binary codes). If you have successfully written your codes inside the text editor, and you select the **verify icon** (the icon that looks like a triangle), this will try to compile the C language inside the editor without necessarily making an effort to send the code to the Arduino IDE. If you enter a word like KENT WAMP inside the Arduino IDE, and then hit the upload icon at the top, you will get a screenshot that look like the one below;

The Arduino has helped you run the word KENT WAMP and the language is not recognizable. This is because the text you entered inside the text editor is not a C language. The error message "KENT WAMP does not name a type" is displayed. In another manner, when you compile a sketch that has no code at all, the compiler will run and tell you that your sketch does not have a loop function. You need to have some boilerplate codes before you can start adding your own codes into your sketch. For Arduino programming, the "boilerplate" code will assume the form of the "setup" and "loop" functions which must be always available in your sketch. So, you must always adapt your sketch to the boilerplate so that your code will run. When you adapt the boilerplate in your code, the compiler will find it acceptable and will bring a

"Done Compiling" message. It will as well tell you the size of the sketch. Let us take a look at the boilerplate code which will later be the starting point for all the sketches that you will be writing in Arduino.

Let us examine the Void setup first. The line void setup() implies that you want to define a function called setup. In Arduino, some functions have been defined for you already, like the digitalWrite and the delay function that you used earlier, whereas you need to define other functions by yourself. The setup and the loop functions are two important functions that you have to define for yourself in every sketch that you write. You have to know that you are not calling the loop and the setup functions just like you called the digitalWrite, but your plan is to create the two functions so that your Arduino can call the functions. To have a good grasp of this concept, see the functions as some set of documents. In the document, you have a part that used fisherman as the shorthand definition for someone that engages in fishing activities. This will enable the document to look concise and straightforward to read than a lengthy and ambiguous one. Functions work in a much similar manner like this. This means you can actually get to define a specific function that you, and even the Arduino program, can deploy anywhere in your sketch. When you look at it very well, you will observe that the two functions (setup and loop) do not actually return any value, so to say, that means you need to say that the two functions are void. Going back to void, these two functions (setup and loop) do not return a value as some functions do, so you have to say that they are void, using the void keyword. Let us say you have a cos function that can perform the cosine trigonometric function of cos, then the cos function will return a value. The value it will return would be the cosine of the angle that was passed to it as the argument. In short, rather than using words to define a function, we simply use functions in C language which we can call anytime to do something for us. The name of the function will usually come after the keyword void followed by a parenthesis to accommodate the argument. A semicolon doesn't follow the closing parenthesis because you are defining a function, and not that you are calling the function. So you must define what should happen when something does call the function. The action that you need to occur when something calls on the function should be put in between the curly braces. The curly braces together with the codes between them are referred to as blocks of code. Note that even though you need to define both the functions setup and loop, you don't necessarily have to put lines of code inside them. But you need to be reminded that if you don't put lines of codes, your sketch might look a bit unattractive.

The Blink Example

The reason why Arduino possesses the two functions setup and loop, obviously, is to separate the things that only need to be done once, when the Arduino starts running its sketch, from the things that have to keep happening continuously. The function setup will just be run once when the sketch starts. Let us add some code to it that will blink the LED built onto the board. Add the lines to your sketch so that it appears as follows and then upload them to your board:

```
void setup()
{
  pinMode(13, OUTPUT);
  digitalWrite(13, HIGH);
}

void loop()
{
}
```

The setup function itself calls two built-in functions, which are the pinMode and digitalWrite. You already have an idea of what digitalWrite is all about, but pinMode might seem new to you. The pinMode function assigns a particular pin to either be an input or an output. So, the process involved in turning the LED on is actually a two stage process. First, you need to set pin 13 (LED) to be an output, and second, you need to assign that output to be high (5V). When you run this sketch, on your board you will observe that the LED comes on and stays on. This is not very exhilarating, so we can at least try to make the LED flash by turning it on and off in the loop function instead of using the setup function. The pinMode call can be put in the setup function since you are only required to call on it once. Although, the project won't get terminated if you move the pinMode into the loop. To get things done, you can modify the sketch above to something like the code below;

```
void setup()
{
  pinMode(13, OUTPUT);
}

void loop()
{
 digitalWrite(13, HIGH);
 delay(500);
 digitalWrite(13, LOW);
}
```

Run the sketch above and see the result. It may not be quite what you were expecting because it appears as if the LED is on all the time. Let us see what is happening; try to prompting the sketch above one-step at a time in your head and you can summarize like below;

1. Run the sketch and set pin 13 to be an output.

2. Run loop and set pin 13 to high (LED on).

3. Delay for half a second.

4. Set pin 13 to low (LED off).

5. Run loop again, going back to step 2, and set pin 13 to high (LED on).

The problem lies in step 4 and step 5. You will notice that the LED is being turned off, and then it is turned on again when step 5 instructs the code to go to step 2 (which clearly says Set pin to high). However, the process occurred so fast that it looks like the LED is turned on all the time. The microcontroller chip on the Arduino can perform 16 million instructions per second. That is not 16 million C language commands, but it is still very fast. Therefore, our LED will only be off for a few millionths of a second. To fix the problem, you need to add another delay after you turn the LED off. Your code should now look like this:

```
// sketch 3-01
void setup()
{
  pinMode(13, OUTPUT);
}

void loop()
{
 digitalWrite(13, HIGH);
 delay(500);
 digitalWrite(13, LOW);
 delay(500);
}
```

Variables

You should also notice that pin 13 was used and referenced in three places. If you use a different pin, you will only need to modify the code in three different places inside the code above. If you want to delay the rate of blinking of the LED, you can change the number 500 to any other number in those places where the number 500 has appeared. When defining a variable in the Arduino language C, you have to indicate the type of the variable. If you want your variables to be whole numbers, which in C are called integers (ints). So if you want to define a variable called ledPin with a value of 13, you have to write the following:

```
Int   ledPin   13;
```

You can observe that ledPin is a name, so just names of functions; you should not add any spaces in between the names. Notice that because ledPin is a name, the same rules apply as those of function names. So, there cannot be any spaces. The norm is to begin variables using a lowercase letter and start each new word using an uppercase letter. Let us fit this into your Blink sketch as follows:

```
int ledPin = 13;
int delayPeriod = 500;

void setup()
{
  pinMode(ledPin, OUTPUT);
}

void loop()
{
 digitalWrite(ledPin, HIGH);
 delay(delayPeriod);
 digitalWrite(ledPin, LOW);
 delay(delayPeriod);
}
```

If you check the sketch above very well, you will notice that another variable called "delayPeriod" has been inserted. The "delayPeriod" has replaced the number 500 in the previous sketch and the ledPin variable has replaced the integer variable 13. If you need to make your sketch blink very fast, you can modify the value of delayPeriod in one place. Try to change the number 500 to, say 100, and then run the sketch on the Arduino Uno board. This is one way of experimenting with variables and having different results.

You can as well modify the sketch such that the blinking of the LED starts very fast and then goes dim or reduced making it look like the Arduino board is getting tired. To carry this out, all that you might need to do is to insert something into the delayPeriod variable every time you do a blink. You can modify your sketch by including the single line at the end of the loop function and then initiate the sketch like the one below;

```
int ledPin = 13;
int delayPeriod = 100;

void setup()
{
  pinMode(ledPin, OUTPUT);
}

void loop()
{
 digitalWrite(ledPin, HIGH);
 delay(delayPeriod);
 digitalWrite(ledPin, LOW);
 delay(delayPeriod);
 delayPeriod = delayPeriod + 100;
}
```

Hit the Reset button and watch it start from a fast rate of flashing again. Your Arduino Uno is now doing some sorts of arithmetic. Every time that loop is called, it will carry out the normal flash of the LED, but then it will add 100 to the delayPeriod variable.

Experiments in C Language

You need a means that you can use to test your experiments in C. One method is to insert the C that you want to test out inside the setup function, evaluate them on the Arduino interface, and then have the Arduino show any output back to something called the Serial Monitor. The Serial Monitor is inside the Arduino IDE. You can have access to the Serial Monitor by tapping on the top right section in the toolbar. The Serial Monitor serves as a channel of communication between the Arduino Uno and your computer. You can enter any message inside the text area at the top part of the Serial Monitor and then hit the send button to send the message to the Arduino Uno. Likewise, the Arduino Uno can relay any message and you will get it on the Serial Monitor. The information, in both cases, is sent through Universal Serial Bus link. The Serial.println is a built-in function that you can deploy in your sketches to send back a message to the Serial Monitor. The Serial.println cannot take more than a single argument, which consists of the information, usually a variable that you intend to send.

Numeric Variables and Arithmetic

The last thing you did was add the following line to your blinking sketch to increase the blinking period steadily:

delayPeriod = delayPeriod + 100;

When you take a closer look at the line above, you will observe that it consists of a variable name, an equal to (=) sign and what you can otherwise refer to as an expression (delayPeriod + 100). The equal to sign is there as an assignment. That implies that it is the one that is assigning a new value to a variable, and the value it is given is a function of what comes right after the equals sign and just before the semicolon. In this instance, the new value to be given to the delayPeriod variable will be the old value of delayPeriod plus 100. We can test out this new mechanism to examine what the Arduino is doing by running the following sketch, and then opening the Serial Monitor:

```
void setup()
{
  Serial.begin(9600);
  int a = 2;
  int b = 2;
  int c = a + b;
  Serial.println(c);
}
void loop()
{}
```

Let us take a slightly more complex example, the formula for changing a temperature from degrees Centigrade into degrees Fahrenheit is to multiply it by 5, divide by 9, and then add 32. So you could write that in a sketch like the one below;

```
void setup()
{
  Serial.begin(9600);
  int degC = 20;
  int degF;
  degF = degC * 9 / 5 + 32;
  Serial.println(degF);
}
void loop()
{}
```

There are some few things you will likely observe here. First, note the line below:

Int degC = 20;

When you see a line like the one above, it is doing two tasks; it is declaring an int variable called degC, and it is saying that its initial value will be 20. Alternatively, you could separate these two things and then come up with one below;

Int degC;

degC = 20;

You should declare any variable that you want to declare just once. However, you can still assign your variable a value as many times as you desire. So, in the Centigrade to Fahrenheit instance, you are declaring the variable degC and assigning it an initial value of 20, but when you declare the degF, it does not get an initial value. Its value gets assigned on the next line, according to the conversion formula, before being sent to the Serial Monitor for you to see.

Commands

The C language used in Arduino has a wide number of built-in commands. In this part, you will be able to see some of these commands and how you can deploy them in your Arduino Sketches.

If command

For all the sketches you have been seeing so far, the basic assumption that was made was such that the lines of code were executed in order just one after the other without exception. But we can have an instance where you don't even want to tow that lane – perhaps you only want to execute only part of your sketches and not everything. How do you get this done?

Let us use the previous example of the LED blinking light which is already slowing down gradually. At the moment, it will start getting slower and slower until each blink is lasting hours. Let us take a look at how we can actually modify it so that once it has slowed down to a certain point, it reverts back to its fast starting speed again.

You can deploy the **if** command to achieve this by modifying the last sketch (where the LED is already slowing down) as follows;

```
int ledPin = 13;
int delayPeriod = 100;

void setup()
{
  pinMode(ledPin, OUTPUT);
}

void loop()
{
  digitalWrite(ledPin, HIGH);
  delay(delayPeriod);
  digitalWrite(ledPin, LOW);
  delay(delayPeriod);
  delayPeriod = delayPeriod + 100;
  if (delayPeriod > 3000)
  {
    delayPeriod = 100;
  }
}
```

The **if** command looks a bit like a function definition, but this resemblance is not deep as it is only on the surface level. The word in the parenthesis is not an argument; it is what is called *a condition*. So in this case, the condition is set such that the variable **delayPeriod** has a value that is more than 3,000. If this is true, then the commands inside the curly braces will be executed. In this case, the code sets the value of **delayPeriod** back to 100. If the condition is not true, then the Arduino will just continue on with the next thing. In this case, there is nothing after the "if", so the Arduino will run the **loop** function again.

If you run these series of events right inside your head, you can have a better grasp of what Arduino is doing. Look at what is happening below

1. Arduino initiates **setup** and prompts the LED pin to be an output.

2. Arduino begins to run the **loop**.

3. The LED light turns on.

4. A delay occurs.

5. The LED turns off.

6. A delay occurs.

7. Add 100 to the **delayPeriod**.

8. If the delay period is greater than 3,000 set it back to 100.

9. Go back to step 2.

We used the symbol <, which implies less than. The less than (<) is a typical example of comparison operator.

You can deploy the double equal to symbol (==) to make comparison between two numbers. Don't confuse the double equal to (==) sign with an ordinary equal to (=) which is only used to assign values to variables.

There is another type of **if** that enables you to carry out one task if the condition is met, and then carry out a different task if the condition is not true.

The *for* command

While you will be able to execute a number of commands under different conditions, you might as well decide to execute a series of commands a number of times in a program. The loop method is a good way of achieving this. The loop function will automatically start again as soon as all the commands in the program have been executed.

However, there are times when you will need more levels of controls than this. Let us, for instance, write a sketch that can blink 20 times, then paused for just 3 seconds, and then start again. You can achieve this by simply repeating the same code over and over again in your **loop** function, like the one below;

```
int ledPin = 13;
int delayPeriod = 100;

void setup()
{
  pinMode(ledPin, OUTPUT);
}

void loop()
{
 digitalWrite(ledPin, HIGH);
 delay(delayPeriod);
 digitalWrite(ledPin, LOW);
 delay(delayPeriod);

 digitalWrite(ledPin, HIGH);
 delay(delayPeriod);
 digitalWrite(ledPin, LOW);
 delay(delayPeriod);

 digitalWrite(ledPin, HIGH);
 delay(delayPeriod);
 digitalWrite(ledPin, LOW);
 delay(delayPeriod);
// repeat the above 4 lines another 17 times

 delay(3000);
}
```

But doing it like the one you have above takes a lot of time entering the codes inside the editor. There are other ways of getting this done by using a **"for loop."**

The sketch to achieve the task above by using a **for loop** is as shown below which is much easier;

```
int ledPin = 13;
int delayPeriod = 100;

void setup()
{
  pinMode(ledPin, OUTPUT);
}

void loop()
{
  for (int i = 0; i < 20; i ++)
  {
   digitalWrite(ledPin, HIGH);
   delay(delayPeriod);
   digitalWrite(ledPin, LOW);
   delay(delayPeriod);
  }
 delay(3000);
}
```

The first item in the parentheses after **for** is the variable declaration. This specifies a variable to be used as a counter variable and assigns it an initial value—in this case, 0.

The second part is a condition that must be essentially true for you to stay in the **loop**. In this case, you will be able to stay in the **loop** as long as the value of **i** is still less than 20, but as soon as **i** is equal to 20 or more, the program will stop carrying out the task inside the loop.

The last part is the action you want to be doing every time you have carried out all the tasks in the **loop**. In this case, that is to raise the value of **i** by 1 so that it will, after 20 trips around the **loop**, stop to be less than 100 and cause the program to exit the **loop**.

Try entering the code and then run it. As an advice, you should endeavor to be typing these codes as this is one way to avoid unnecessary punctuations and errors, which will not give you results.

One potential disadvantage of this method is that the **loop** function will take a long time. This is not an issue for this sketch, since all that it is doing is to flash the LED. But often, the **loop** function in a sketch will also be checking that keys have been pressed or that serial communications have been received. The processor will not be able to check for any serial communication if it is busy inside a **for loop.** This is why it is necessary to let the **loop** function to run as fast as it can so that it can be run as frequently as possible.

The sketch below shows you how to get this done:

```
int ledPin = 13;
int delayPeriod = 100;
int count = 0;
void setup()
{
  pinMode(ledPin, OUTPUT);
}

void loop()
{
 digitalWrite(ledPin, HIGH);
 delay(delayPeriod);
 digitalWrite(ledPin, LOW);
 delay(delayPeriod);
 count ++;
 if (count == 20)
 {
   count = 0;
   delay(3000);
 }
}
```

You may have noticed the following line:

Count ++;

This is just C shorthand for the following:

Count =count + 1 ;

So now each time that **loop** is run, it will take just a bit more than 200 milliseconds, unless it is the 20th time round the loop, in which case it will take the same plus the three seconds delay between each batch of 20 flashes.

While loop

Another method of looping in C is to deploy the **while** command to replace the "**for**" command. You can also do the same thing as the preceding **for** example when you use a **while** command as follows:

```
int i = 0;
while (i < 20)
{
  digitalWrite(ledPin, HIGH);
  delay(delayPeriod);
  digitalWrite(ledPin, LOW);
  delay(delayPeriod);
  i ++;
}
```

The expression in parentheses after **while** must be true if it wants to stay in the **loop**. When it is no longer true, then the sketch continues running the commands after the final curly brace.

CHAPTER FIVE

INTERFACING ARDUINO WITH PYTHON PROGRAMMING LANGUAGE

In chapter one, we discussed the nitty-gritty of the python programming language and how you can write a script with the language. In this chapter, we will take a look at how we can bridge the python language with Arduino and then control some basic hardware connected to the Arduino through the python programming language.

Two major platforms can be used to bridge Arduino with python. These include;

- **The Arduino Firmata protocol**
- **The python's serial library (pySerial)**

Most times, people used the Arduino Firmata protocol to interface Arduino with python, but the Firmata protocol can also be used to develop a number of useful applications.

Now that you have decided to get your hands dirty with some hardware that can be connected to Arduino, you are expected to get ready some of these components;

- **A breadboard**
- **An LED**
- **USB cable**
- **Arduino Uno** (there are other variants of Arduino, but the one for these projects is the Uno)
- **A resistor**

Connecting Arduino with your computer

In chapter two, you were able to use codes to successfully communicate with the Arduino board using the Arduino IDE on your computer. Nonetheless, if you were not able to communicate successfully with these codes, this section is still for you to get things done and connect your Arduino board with your computer using a USB. The first time you are to carry out is to connect your computer with the Arduino board using a Universal Serial Bus (USB) cable and then follow these instructions to connect Arduino based on the operating system that your computer is running.

Linux environment

You need to ensure that you have the upgraded version of the Ubuntu Linux. As soon as you connect your Arduino board to the computer, you will be greeted with a prompt on the Arduino IDE telling you to add your username to the dialout group. Note that the username here is your Linux username. If you don't know your username, quickly open the terminal window and type **whoami** or you can alternatively tap the system menu in the top right corner of the screen and look for your username at the bottom entry of the drop-down menu. Tap the **"Add"** button on the prompt to add your username which you have already obtained and then log out from your system. The changes will take effect automatically without necessarily having to restart your device. Log in to the interface with the same username you entered earlier (remember to choose a username that you can remember) and then enter the Arduino IDE on the Ardunino software. If you don't get the prompt to add your username (as some people on various online platforms complained of not getting this prompt on their Linux), then you need to do a little troubleshooting. Kindly check the **Serial port** menu from the **Tools** menu of your Arduino IDE. It is possible that you might have installed some other software on your computer that has already added your username to the dialout group. If you couldn't see the **Serial Port** option on the Arduino IDE, then try and execute the prompt (in the terminal) below;

$ sudo usermod -a -G dialout <username>

Note that the username here is your Linux username.

The script you add above will add your username to the dialout group, and this will work for other Linux models. The Arduino board in Linux will usually connect as **/dev/ttyACMx** where the "x" here is just an integer value which depends essentially on the physical port address of your computer.

If you are using the Fedora Linux distribution, simply add the uucp and lock groups with the dialout group in order to control the serial port:

$ sudo usermod -a -G uucp,dialout,lock <username>

Mac OS X

The Mac OS X configures your Arduino Uno board as a network interface anytime you connect the board to the system through a serial port. If you have the Mac OS X Mavericks, navigate to **Network** from **System preferences** once you connected your Arduino. You will get a prompt telling you that you have connected a new network interface. Tap on **ok** for **Thunderbolt Bridge** and then select **apply.**

If you have the Mac OS X Lion or the later models, you will get a prompt (after connecting the Arduino board to your computer) that will tell you to add a new network interface. But unlike the Mac OS X Mavericks, you will not navigate to the network preferences on your computer. Once you notice the network interface that reads **not connected** highlighted in red, you should not fear as it is just a prompt and it doesn't determine anything. Enter your Arduino IDE and scroll to **Serial Port** from the **Tools** option. Note that the serial port on which your Arduino board has been connected might vary based on your OS X version and the physical port to which it is connected. Ensure that you choose a **tty** interface for a USB modem.

Windows

The configuration of your Arduino serial port is not so difficult if you have a Windows device. The Windows operating system will help you to install all the necessary drivers the very first time you slot your Arduino Uno board inside the computer. Once the necessary installations have been completed, choose the appropriate COM port right from the serial port of the Arduino menu bar. Tap on **Tools** and then select the **Serial port** option in the menu bar to choose your COM port.

You might follow all the steps highlighted above for the various operating systems, ranging from Linux, Mac and Windows, and still don't get to see the **Serial port.** If this is the case with you, then a little troubleshooting is necessary. The problem can be caused by two major reasons: either the serial port is currently being used by another software on your computer or your Arduino USB drivers have not been installed properly. If you observe that the serial port is currently being used by any other software aside from the Arduino IDE, kindly terminate the software and then restart the Arduino IDE. At times in Linux, the **bratty** library might not go well with the Arduino serial interface. Simply remove this library, log out of the IDE and then log in again:

$ sudo apt-get remove brltty

If you are using Windows, you can try to reinstall your Arduino IDE as this will help to install and configure the Arduino Uno USB drivers again.

The Firmata protocol

Arduino is an inbuilt hardware gadget with I/O pins that can communicate with an external device. You can write a sketch that can control these external devices attached to the Arduino Uno right inside the Arduino IDE. And just like magic, you can see your devices (Arduino) receiving communications from the computer. If it is possible to write an Arduino program capable of transferring the full control of the Arduino I/O pin to another software, then you don't need to stress yourself writing an Arduino for every change you want to effect in Arduino. This can be solved by writing such an Arduino program that can be controlled with the serial port. The **Firmata protocol** is the platform of choice that can handle this.

Firmata is a broad protocol that enables communication between the Arduino microcontroller and the software hosted on your computer. If you have a software that has been hosted on your computer that can perform serial communication with the microcontroller can actually do this with the help of Firmata. With Firmata, you don't have to start uploading and modifying any sketch before your software gets direct and full access to the Arduino.

To use the Firmata protocol, you as a developer can upload a sketch that supports the Firmata protocol straight to the Arduino client as a one time process. By so doing, you can have the chance to start writing custom software on the host computer and then perform hard tasks on it. The custom software will be able to give commands to the Arduino Uno board equipped with Firmata by using a serial port. The art of writing custom Arduino sketches still holds when it comes to standalone applications where the Arduino board has to carry out some local tasks.

Uploading the Firmata sketch to the Arduino board

One of the ways - perhaps the best way- that you can deploy to test the Firmata protocol is to upload a standard Firmata program to your Arduino Uno board and then use the testing software from the host. You will get to upload an Arduino sketch which already features the standard Firmata program to the board. If you want to get the best out of Firmata, you need the latest version of the Firmata firmware and you don't need to stress yourself to get it. If you have the latest and upgraded Arduino IDE, then you are sure it already has the latest model of the Firmata firmware. The methods below show you exactly how you can upload the Firmata program to our Arduino Uno board;

- Open the Arduino sketch interface on your computer. Navigate to **File,** and then select **Example.** Scroll down from the **example** menu and choose **Firmata.** Scroll down one more time and select **standard Firmata.**

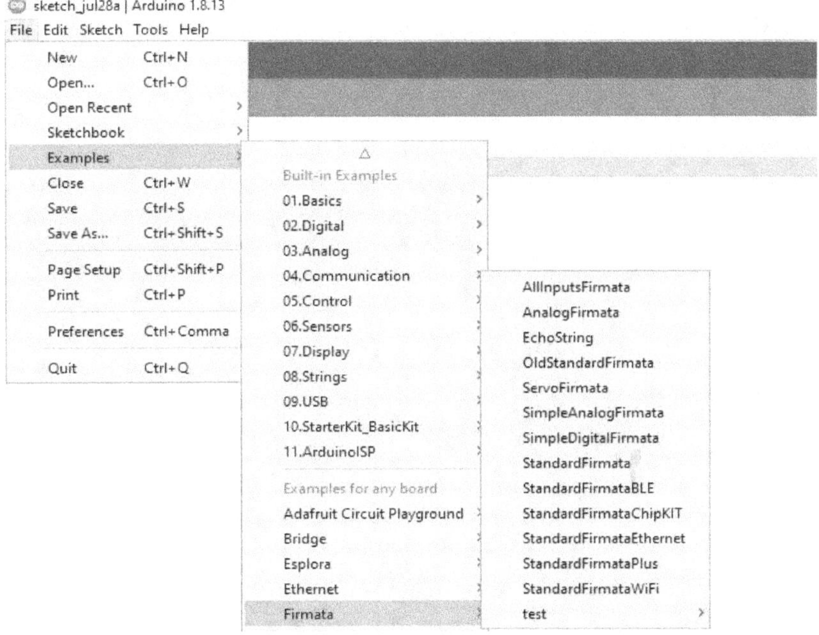

- Once you tap on the **StandardFirmata,** you will be greeted with a new window with a new sketch space loaded in the editor. It is important to tell you not to change anything in the displayed window; this is because the test software will not work any longer if you try to modify anything within the code.

- Once you have successfully opened the **Standard Firmata** sketch, the next thing is to compile it for your Arduino board. You have already seen how to connect the Arduino Uno board to your computer and choose the right serial port. But if you noticed the new sketch space opened for you has a different configuration from that, kindly navigate to **Tools** and then scroll down to choose the right Arduino board and port.

- To compile the current sketch, tap on the **Verify** icon from the toolbar as shown in the screenshot below. You can as well compile the current sketch by selecting **Sketch | Verify/ Compile** or tapping on *Ctrl + R* (*command + R* if you have the Mac OS X):

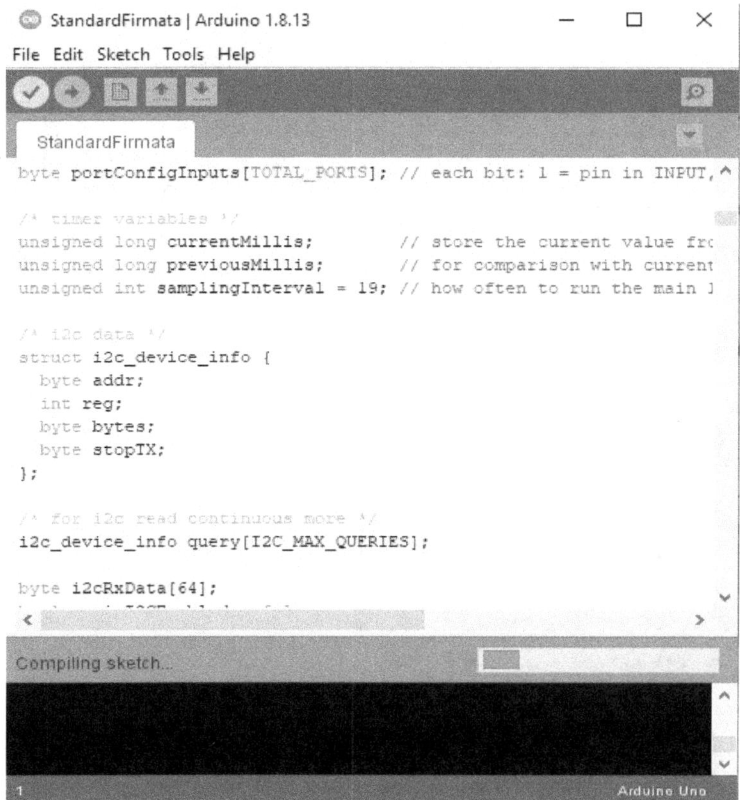

- The standard Firmata sketch above is already compiling (see the bottom left corner of the screenshot) by selecting the **compile** icon in the toolbar section. By pressing the upload button or compile button, you can upload the codes displayed in the editor above straight to the Arduino Uno board.

- Once the code is done compiling, you will get a notification at the bottom of the window like the one in the screenshot below;

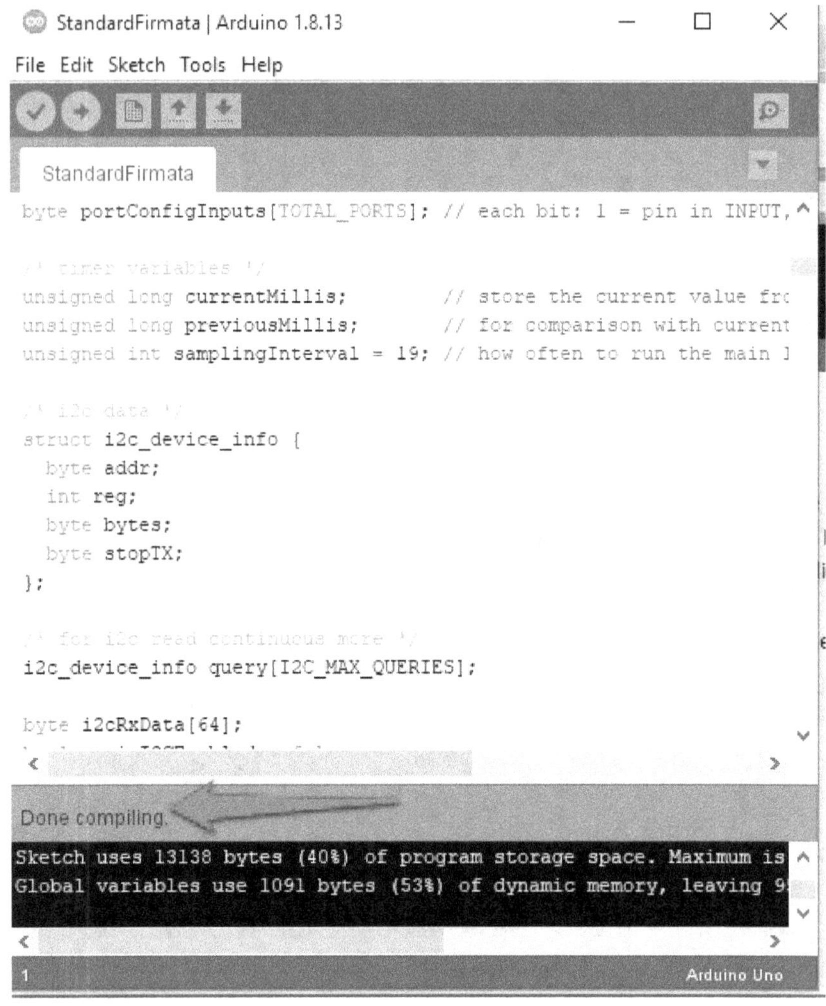

After successful compilation of the codes, your Arduino Uno board is now available with the latest Firmata firmware. The next thing is to start testing the Firmata protocol.

Testing the Firmata protocol

You can remember how you once used an on-board (within the Arduino board) LED light at pin 13 to test the **blink program** by writing a script for the blinking process, now is the time to get you started with how to assemble hardware parts using the Arduino Uno board. Although, this time we are not going to use an on-board LED but rather an external LED which will be connected to your Arduino. To get started, you will need a LED, 1 kilo ohm resistor and an Arduino board. First, you need to disconnect your Arduino board from the computer to avoid damage due to static electricity. The LED we are talking about comes with two legs; the short leg is grounded with the 1k-ohm resistor and the long leg will be connected to the pin 13 of the Arduino Uno board. Just something that will look like the image below;

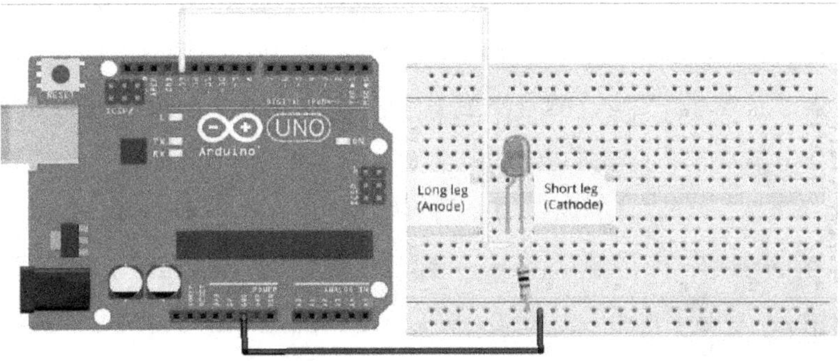

The attached LED above will be deployed to test some basic functions of the Firmata protocol. The Firmata code has already been uploaded to the Arduino and you can now get ready to start controlling the LED attached with your computer (host computer).

There are many ways you can communicate with the Arduino board from your computer using the Firmata protocol. One of such ways is to write your own program in python by using the supported library or with the prebuilt testing software.

One free tool you can use for this kind of communication is the Firmata firmware. You can also get many test tools from the Firmata's official website at http://www.firmata.org . Follow the steps below to test the deployment of the Firmata protocol;

1. Depending on the operating system on your computer, you are required to download the right version of the firmata_test program that fits your computer.

2. After you have successfully downloaded the right Firmata program, attach your Arduino board (which has already been joined to the LED) to the host computer by using the USB cable and then run the downloaded firmata_test program. When the program has been executed successfully, you will be greeted with a window telling you that your program has been executed successfully.

3. This step should not involve error on your side to avoid your program nor running successfully, navigate to the top section of the Firmata test program and select **port.** From the **port** section, choose the right port from the port dropdown list. You need to be sure that you are selecting the same port that you used when you wanted to upload the Arduino sketch. You also need to ensure that the Arduino IDE is not connected to the Arduino board with the same port number you are choosing on the Firmata test program. This is because the serial port can only give access to just one program at a time.

4. Once you have successfully selected the Arduino serial port, the program will bring multiple drop-down boxes including buttons with labels that have the pin number. You

can already observe from the image below that the Firmata program has been preloaded with 12 digital pins (pin 2 to pin 13) and six (6) analog pins (pin 14 to pin 19).

As you start using your Arduino Uno board for your applications, the Firmata test program will only load pins that are part of Arduino Uno. If you have, let us say the Arduino Mega or any other board, the number of pins displayed in the Firmata program will be according to the pins supported by that particular variant of the Arduino board that you are using.

Working with the firmata_test program on Linux

If you have a Linux platform, you might have to change the property of the downloaded Firmata test file to make it executable. From the same directory, prompt the command below in the terminal to make it executable:

$ chmod +x firmata_test

Once you have modified the permissions, you can then use the following command to run the program right from the terminal:

$ ./firmata_test

5. As you can verify in the program window below, you have two other columns including the column that contains the labels. The second column in the program enables you to indicate the role for the appropriate pins. You can indicate the role of digital pins (in the case of Arduino Uno, from 2 to 13) as input or output. As shown in the screenshot below, you will be able to see **Low** in the third column immediately you

choose the role of pins 2 and 3 as input pins. This is right, since we don't have any input connected to these pins. You can do some other manipulations with the program by changing the functions and values of multiple pins.

You have connected the LED to digital pin 13, and you are not expected to do any physical changes on the board while you are still working around with the other pins.

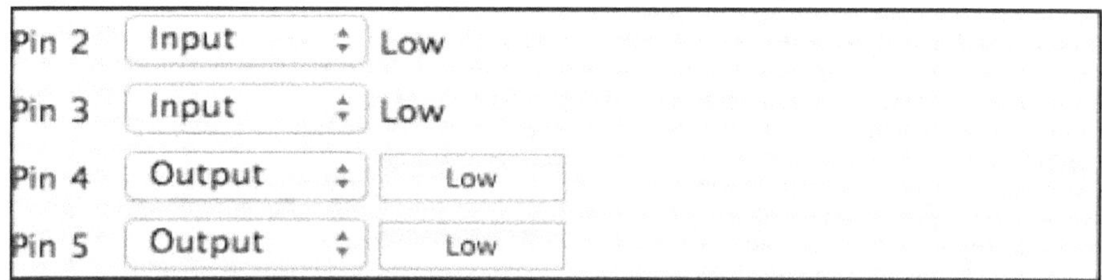

6. You can now choose pin 13 as the output pin and hit the **Low** button. This will modify the

Button's label to **High** and you will notice that the LED is turned on. By carrying out this process, you have modified the logic of the digital pin 13 to 1, which implies **High**, which translates to +5 volts at the pin. This potential difference will be good enough to light up the LED. You can as well change the level of pin 13 back to 0 by tapping on the button again and then turning it to **Low**. This will modify the potential difference back to 0 volts. The Firmata program that you have deployed here is good for testing fundamental things, but you cannot use it to compute complex applications by using the Firmata protocol. When you are in a real world scenario, you will need to execute this Firmata method by using custom mode, which can as well implement smart logic and algorithms apart from switching the LED.

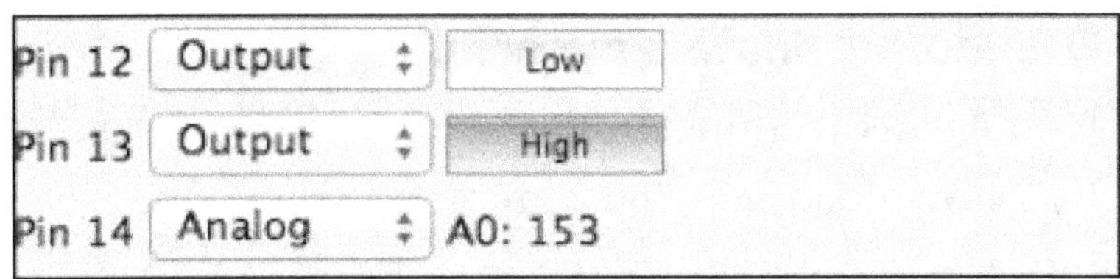

The pySerial Library

You don't necessarily need to write your own library by specifying and implementing functions before you can allow communication on a serial protocol. This is because it is time consuming and a little bit inconvenient. This stress can be avoided by using the python library called the pySerial. The pySerial library allows communication with your Arduino Uno by encapsulating the access for the serial port. This module will provide access to the serial port settings through the Python properties and enables you to configure the serial port directly through the interpreter. The pySerial will be the bridge for any further future communication between the Python and your Arduino Uno. Let us get started by installing pySerial;

- Open your terminal or the command prompt and then prompt the command below:

> **pip install pyserial**

If you are using the Windows operating system, you do not need any administrative level user access to prompt the command above, but you do require root privileges before you will be able to install python applications if you are using a Unix-based operating system. Just prompt the command below for Unix;

$ sudo pip install pyserial

You can as well install the pySerial library from source by downloading the archive from the web at http://pypi.python.org/pypi/pyserial, unpack the archive, and then run the command below from the pySerial directory;

$ sudo python setup.py install

- If Python and Setuptools have been properly installed, you will see the output below at the command line after the installation is complete:

.. Processing dependencies for pyserial

Finished processing dependencies for pyserial

This means that you have successfully installed the pySerial library and you are good to start programming your Arduino with pySerial library.

- Now, to check whether or not pySerial is successfully installed, start your Python interpreter and import the pySerial library using the following command:

>>> import serial

Working with the pySerial

Previously, you have seen how you can use the **Standard Firmata** to sketch on your Arduino. Now, you will be seeing how you can use another simple Arduino sketch that is capable of implementing serial communication which can be captured on the python interpreter.

- From the Arduino IDE, scroll to the **File** menu and tap on **Example.** Then select **Basic >> DigitalReadSerial.**

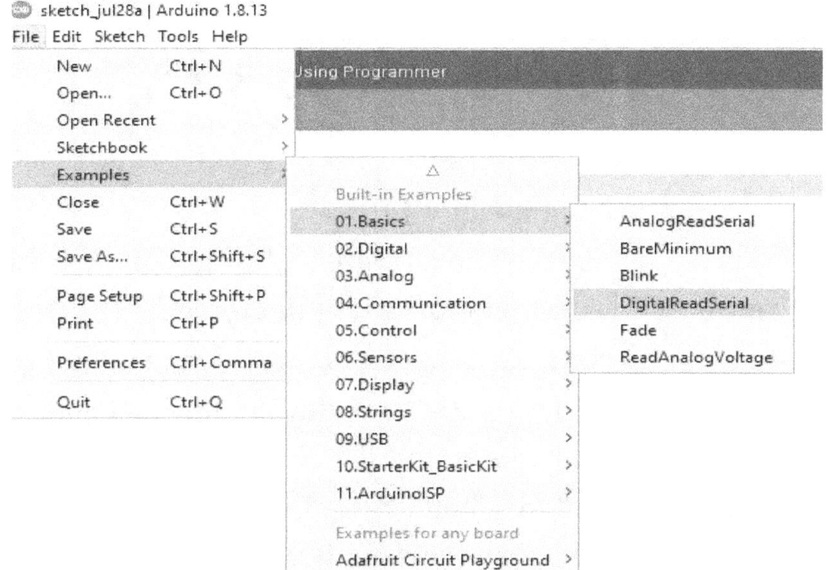

- Simply compile and then upload the DigitalReadSerial program to the Arduino board by tapping the DigitalReadSerial. You will be greeted with a window containing the codes. Select the verify icon (icon that looks like a v-shape) at the top to start compiling the code. Also, choose the appropriate serial port, which must be the same port as the one on which your Arduino is connected by selecting the **Tools** option at the top section. Note down the port name. This Arduino code reads a digital input on pin 2, and then prints the result to the Serial Monitor. While the Arduino Uno board is still connected to your computer, open your python interpreter. And then prompt the command below on the python interpreter. Ensure that you replace the /dev/ttyACM0 part of the code below with the port name that you noted down previously.

```
>>> import serial
```

```
>>> s =
serial.Serial('/dev/ttyACM0',9600)
>>> while True:
    print s.readline()
```

- When you execute the prompt above, you will have repeated 0 values in the Python interpreter. Hit *Ctrl +C* to stop the code. As you can observe, the Arduino code will keep sending messages due to the fact that you used a loop function in the sketch.

In the preceding Python script, the serial.Serial method interfaces and opens the specified serial port, while the readline() method reads each line from this interface, terminated with \n, that is, the newline character.

Bridging pySerial and Firmata

If you can remember when we treated the Firmata protocol, you have already seen how useful it is to use the Firmata protocol rather than constantly modifying the Arduino sketch and uploading it anytime you are running some simple programs.

The pySerial is just a simple library that initiates a bridge between Arduino and Python through a serial port, but it doesn't have any support for the Firmata protocol. Python has an advantage for the fact that it has a library for almost every package. And as such, there is a python library called pyFirmata built on pySerial to give necessary support to the Firmata protocol. There exists some other Python libraries that also give support to Firmata, but here we will only be focusing our attention on pyFirmata.

1.The first thing is to install the pyFirmata, just the way you used to install any other python package, by deploying the setup tools;

$ sudo pin install pyfirmata

You will still remember that in the previous section, while testing the pySerial, you uploaded the DigitalSerialRead sketch to the Arduino board.

2. To communicate using the Firmata protocol, you will need to upload the **StandardFirmata** sketch again, just as you did in the *Uploading a Firmata sketch to the Arduino board* section.

3. Once you have uploaded the **StandardFirmata** sketch, open the Python interpreter and prompt the script below. The function of the script is to import the pyFirmata library to the interpreter. The script also defines the pin number and the port.

>>> import pyfirmata

>>> pin= 13

>>> port = '/dev/ttyACM0'

4. The next step is to link the port with the board type of the microcontroller.

>>> board = pyfirmata.Arduino(port)

In the process of executing the previous script, you observed that two LEDs on your Arduino Uno will be on as the interaction between the python interpreter and the Arduino board is established.

You can still remember that in the *Testing the Firmata protocol* section, you used a prebuilt program to turn an LED on and off. You can carry out these functions directly from the prompt once the Arduino Uno board has been properly associated with the python interpreter.

5. You can now start playing with Arduino pins. Turn on the LED by executing the following command:

```
>>> board.digital[pin].write(1)
```

6. You can turn off the LED by executing the following command. Here, in both commands, we set the state of digital pin 13 by passing values 1 (**High**) or 0 (**Low**):

```
>>> board.digital[pin].write(0)
```

7. Similarly, you can as well get to read the status of a pin from the prompt.

```
>>> board.digital[pin].read()
```

CHAPTER SIX

ARDUINO INPUT AND OUTPUT

This chapter will take care of how to attach electronics to the Arduino board. Outputs may be digital, which literally connotes switched between being at 0V or at 5V, or it can be analog, which enables users to set the voltage to their desired voltage between 0V and 5V—although it is not necessarily as easy as it sounds, as you shall soon see. In the same vein, inputs can as well be digital (for instance, determining whether a button pressed or not) or analog (such as from a light sensor).

Digital Output: Measuring output with a multimeter

For this work, you will need a multimeter, an LED, the Arduino board and some wires. The Arduino Uno has many pins as it has been discussed previously, and for this experiment you will be using digital pin 4. Attach some wire to the multimeter lead to monitor what is happening with the LED, and then connect the multimeter to the right pin (pin 4) on the Arduino board. The arrangement is as shown below;

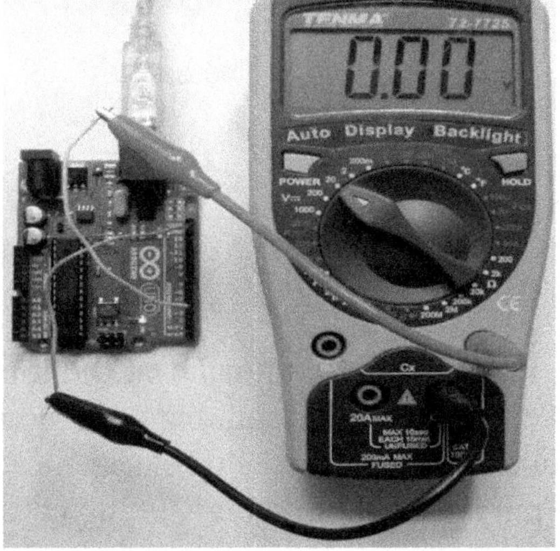

You should set the multimeter to its direct current value of about 0-20V. The negative lead (black) should be joined to the ground pin and the positive lead should be connected to the digital pin at 4 (D4). The wire is connected to the probe

and then inserted into the socket headers of your Arduino
board.

Now, navigate to your Arduino IDE and load the sketch
below;

```
int outPin = 4;

void setup()
{
  pinMode(outPin, OUTPUT);
  Serial.begin(9600);
  Serial.println("Enter 1 or 0");
}

void loop()
{
  if (Serial.available() > 0)
  {
    char ch = Serial.read();
    if (ch == '1')
    {
      digitalWrite(outPin, HIGH);
    }
    else if (ch == '0')
    {
      digitalWrite(outPin, LOW);
    }
  }
}
```

You can see the **pinMode** command at the top of the script
above. What the **pinMode** command is doing is that it
enables Arduino bto configure any electronics connected to
that specific pin to be either the output or an input. See
below;

```
pinMode(outPin, OUTPUT);
```

As you might have understood that the **pinMode** is a built-in
function. The **pinMode** first argument is the pin number
(which is an integer) and its second argument is the mode

which should be either an output or input. The name of the mode must be all capital letters.

This **loop** is set to wait for a command of either **1** or **0** to come from your computer's serial monitor. If the serial monitor inputs a **1,** then the pin 4 (multimeter) will be turned on. If it is not a **1,** the pin 4 will be turned off.

So now that the multimeter has been turned on and plugged into the Arduino, you will be able to see its reading change between 0V and about 5V as you send commands to the board from

the Serial Monitor by either pressing **1** and then **Return** or pressing **0** and then **Return**. The picture below shows a multimeter reading after a **1** has been sent to it from the serial monitor;

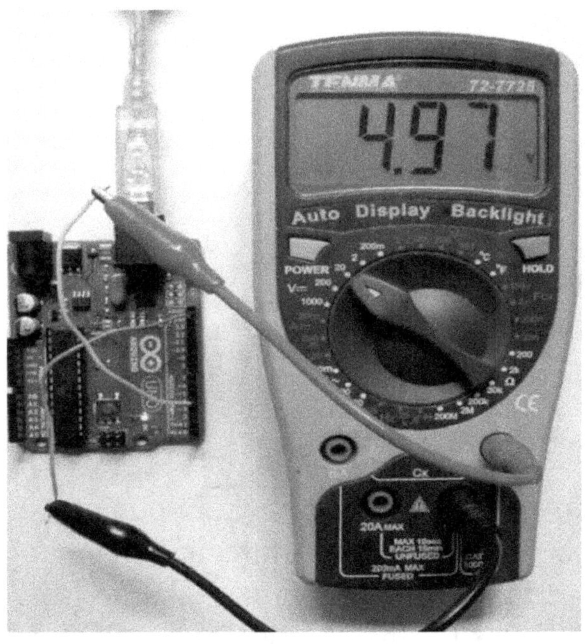

If you don't have enough "D" labeled pins, you can actually use the pins that are labeled "A". Though, the pins labeled "A" are for analog, and you only need to do a little modification by adding **14** to the analog pin number. To do this, simply modify the first line in the above sketch and type pin as **14** instead of **4.** Then move the positive lead of the multimeter to pin A0 on your Arduino Uno.

Digital Input:

You can use a digital input to understand whether a switch has been closed or not. A digital input can either be on or off. If the value of the potential difference at the digital input is less than 2.5V, it will be 0 (meaning off) and if it is above 2.5, it will be 1 (meaning on).

Upload the sketch below onto the Arduino board while disconnecting the multimeter you used previously.

```
int inputPin = 5;

void setup()
{
  pinMode(inputPin, INPUT);
  Serial.begin(9600);
}

void loop()
{
  int reading = digitalRead(inputPin);
  Serial.println(reading);
  delay(1000);
}
```

As it is when you are working with an output, you will need to inform the Arduino in the **setup** function that you will be using a pin as the input. You will be able to get the value of digital input by using the **digitalRead** command. This will return a 0 or a 1.
- Take a small piece of wire and push one end into the D5 socket and then pinch the other end of the wire between your fingers as shown below. When you continue

pinching, you will observe a mixture of ones and zeros showing up on your serial monitor. This is because your fingers are now acting like an antenna picking up electrical signals.

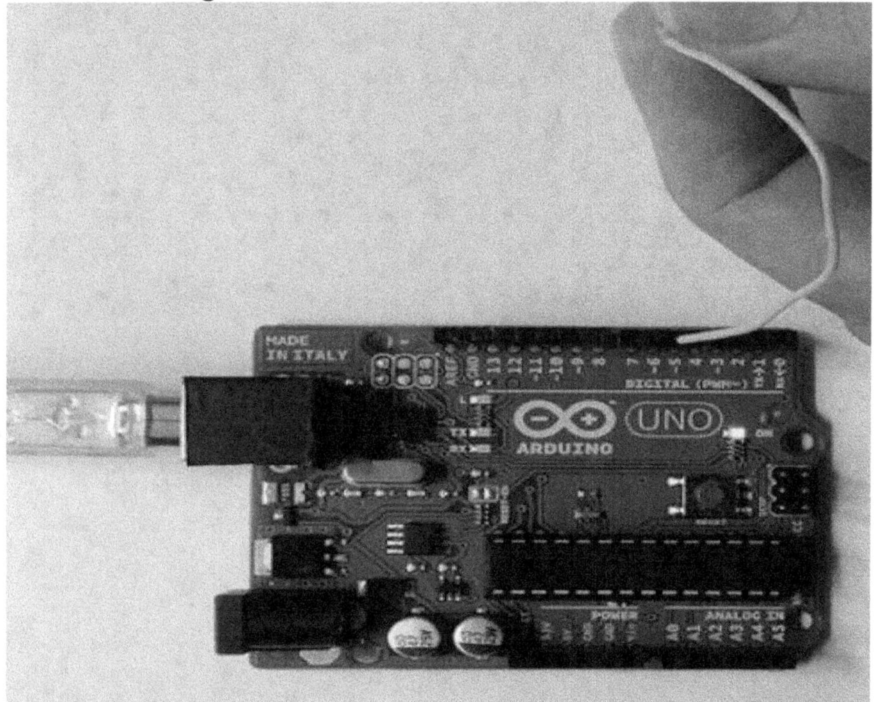

- If you can take the end of the wire you were holding in the image above, and then connect it to the Arduino socket for +5V, then you will start seeing the texts that were a mixture of zeros and ones before now appearing as only ones. See the arrangement below;

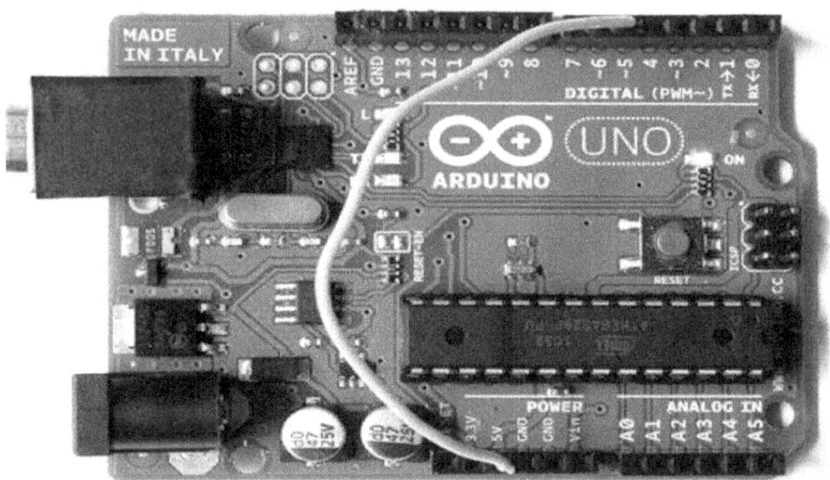

- Now if you can take the end that was in +5V, and then put it into one of the GND connections on the Arduino, you will have your serial monitor displaying zeros.

CHAPTER SEVEN

GETTING YOUR HANDS DIRTY

Building Robots with Arduino

In this chapter, you will learn some basic concepts about how to interface with the Arduino to have a fully functioning robot. There are many things that go into making a robot like; wiring and wire management, order of operation, component placement and arrangement etc. You can begin by getting a commercially available toolkit for building your robot. The kit will contain all the hardware needed to start building robots. In times past, it was almost a herculean task involved in building robots due to the large number of parts you need to build or buy before you can have a fully functioning robot. Now, anybody can build a robot; all thanks to Arduino and readymade kits.

Materials needed to get started

Actually, there is no one-fits-all rule that guides the choice of materials for building an Arduino robot. This is because your choice of materials depends largely on the kind of robot you want to build, and most especially, your budget. Nonetheless, you can still consider the following basic materials that might be handy when building your Arduino robot;

- Any Arduino board: might be an Arduino Uno, Arduino Mini or Arduino micro. Just any.

- Motor driver: the current repository of the robot components. Provides every part with current for propelling.
- Motors: you can easily guess what this one is used for. It allows free rotation of the wheel which makes the robot to easily move around.
- Ultrasonic Distance Sensor: the robot can easily use this to know when there is an obstacle so that it can avoid the obstacle.
- Connectivity module, Bluetooth etc.
- Universal Serial Bus cable: for easy uploading of software
- Vehicle kit or robotic platform: this is like the robot's body. For mounting of various parts.

The list is actually very much and you might not necessarily exhaust it. But the simplest Arduino robots are usually made using most or all of the basic materials outlined above. To ease the stress of having to buy all of these components separately, you can actually buy Arduino **robot kits**. The **Arduino robot kit** is a readymade kit for you that gives you basic gadgets necessary to build your robot. Some examples of readymade Arduino kits that you can buy include; **Arduino robot kits for beginners, Arduino line follower robot kit, AlphaBot2 robot building kits for Arduino etc.**

Let us take a look at one example of a robot built with Arduino

The Arduino Rover robot

Note: This project has been carried out and is solely referenced for the purpose of learning how various hardware and software can make an Arduino robot.

Materials needed

- **Hardware part:** Arduino 101 & Genuino 101, Skeleton Bot – 4WD Mobile Robotic Platform (which contain DC motor etc), Grove starter kit for Arduino (consisting LCD RGB Backlight, sound sensor, buzzer, smart relay, touch sensor, LED etc) and Grove 12C motor driver board (consisting the 12C motor). You can buy the Grove starter kit for Arduino and the Grove 12C motor driver board from online stores.
- **Software part:** Arduino IDE and Nordic semiconductor BLE Toolbox.

Project Achievement

- The built-in wireless Bluetooth Low energy (BLE) communication coupled with a 6 axis Inertial measurement unit (IMU) – for motion.
- The Grove starter kits – adds sensor and remove the need to use a breadboard

Step 1: Assembling the rolling rover chassis

You don't need to worry about getting to assemble the rover kits together as they come pre-assembled when you buy one.

The assembling process literally involves mounting of wheels.

Step 2: Assembling the electrical part

The DC motor, as part of the Skeleton Bot – 4WD Mobile Robotic Platform, has wires already attached to it. The wire on the left motor was attached to motor driver 1 and the wire on the right motor was attached to motor driver 2. The kit will also come with the power cable that was used to connect a battery to the rover. The lithium polymer battery was used in this project. It contains 2S batteries each rated at 3.7V. This gives a 7.4V input for the motor driver.

Talking about output, the Grove I2C motor driver serves to drive the two DC motors and to independently direct their directions. The 12C motor driver also features a 5V regulator which can be deployed to power your Arduino through the 12C bus.

Step 3: Testing the motor control

The wiring was checked for right connection using some basic Arduino codes that spin the four motors at varying speed and direction. The script for everything that was done will be given at the end of the section.

Step 4: Building the rover remote control.

The entire logic involved in building the Arduino rover remote control has been covered on the instructable website at https://www.instructables.com/id/Arduino-101-BLE-Rover-Remote-Control/.

Step 4: programming the Arduino for a "base model" rover.

Starting with a "base model' rover code that contains;

- Rover control code that was used to translate commands into actions that will be carried out by the motor.
- Bluetooth Low Energy communications code used to connect the smartphone to the Arduino interface.

The code was structured to;

- features libraries and then declare global variables

- Setup() function – for code that needed to be run just once run.

- Loop() function – for code that needed to be run repeatedly

- User defined functions – for long codes in a loop.

The Base model code actually featured two functions that were used to control the DC motors;

For communication, the CallbackLED sample which shows the use of Bluetooth Low Energy on the Arduino was modified. This modification involved defining the UART profile that the nRF Toolbox application will like to interact with. The nRF Toolbox application is the one that was used to define the remote control on the smartphone. See the sections of codes that was used below;

Section 1 (include and declare):

- include the wire.h library to start the I2C bus

- include the CurieBLE.h library to allow Bluetooth LE communications

- declare variables defining D2 for the Blue LED, state to hold the RoverControl command char, vSpeedSet to hold the char for speed (0-100)

- make a BLEPeripheral instance, make a BLEService for the UART profile, and then create a rx and tx characteristics for the service

Section 2 (setup function):

- start the I2C bus and the serial link

- define D2 as the output – the D2 is the Grove LED pin.
- Initialize many items related to BLE – LocalName, include the tx and rx feature to the BLE service, specify event handlers for connect, disconnect and rxCharacteristic written, then begin to advertise the BLE service

- start the DC motors as "off" (both + and – pins low for both motors), DC motor speed set at 50%

Section 3 (loop function):

The below image contains all 4 lines of code in the loop() function:

```
blePeripheral.poll();    // check if the rxCharacteristic is written

if(state != prev_state) {   // check if the value from the remote control is new?

RoverControl(state);   // function to respond to control state changes - direction and/or speed

prev_state = state;   // keep a little history
```

Section 4 (user defined functions):

- motor control: MotorSetSpeedAB, MotorSetDirection

- BLEeventhandlers: blePeripheralConnectHandler, blePeripheralDisconnectHandler, rxCharacteristicWritten.

- Rover "state machine" to parse and respond to commands: RoverControl

The Bluetooth Low Energy connected handler turns on the LED when it is connected and the Bluetooth Low Energy will also disconnect when the handler turns off the LED when it is disconnected. **All the codes for this work can be found below;**

The first part of the code contains information about the developer and the licensing information. You can redistribute the code as the developer has made it free.

```
/*
 *
 Arduino 101 based 4WD Rover code developed utilizing:
 * - an Arduino 101 microcontroller board - using the Bluetooth LE radio (BLE), and
 the 6-axis Inertial Measurement Unit (IMU)
 * - motors and chassis from Skeleton Bot - 4WD hercules mobile robotic platform from
 Seeedstudio - early version
 * - Grove I2C Motor Driver Board
 * - Grove Starter Kit for Arduino from Seeedstudio - Grove Button, Grove I2C RGB LCD
 Display, Grove LED Socket,
 * - HC-SR04 Ultrasonic sensor, wired via a small breadboard
 *
 * Most of this project's code is derived from other Arduino samples found
 * on www.instructables.com and www.github.com
 *
 * Much respect and appreciation for the "Makers" who have come before me
 * Lots of knowledge gained from their code and hope my example here is useful to
 others
 *
 * the integration and other bits of original code were written by:
 * Author: Dave Shade
 *
 * It is available for use under the GNU Lesser General Public License
 * see below for details
 *
 * *** Significant pieces of code used from the following: ***
 *
 * Code included for the Grove I2C Motor Driver from:
 * Grove - I2C motor driver demo v1.0
 * by: http://www.seeedstudio.com
 *
 * Code for the RoverControl function taken from the Basic_Robt.ino Sketch:
 * http://www.instructables.com/id/Smartphone-Controlled-Arduino-Rover/
 * by deba168
```

```
 * Code for the Crash Detection derived from the Arduino 101 sample: ShockDetect
 *    Copyright (c) 2015 Intel Corporation.  All rights reserved.
 * Code for the BlueTooth Remote Control derived from the Arduino 101 sample:
 CallbackLED
 *    Copyright (c) 2015 Intel Corporation.  All rights reserved.
 *
 * Specifics for the BLE UART characteristics - to enable the use of the Nordic
 Semiconductor UART applicaion
 * found at: https://www.nordicsemi.com/eng/Products/Nordic-mobile-Apps/nRF-UART-App2
 *
 * Code included from various Grove examples like:
 * Grove RGB LCD display, Grove Button, Grove LED and they are:
 * 2013 Copyright (c) Seeed Technology Inc.  All right reserved.
 *
 * All code utilized is covered under the license agreement described below:
 *
 *  This demo code is free software; you can redistribute it and/or
 *  modify it under the terms of the GNU Lesser General Public
 *  License as published by the Free Software Foundation; either
 *  version 2.1 of the License, or (at your option) any later version.
```

```cpp
 * This library is distributed in the hope that it will be useful,
 * but WITHOUT ANY WARRANTY; without even the implied warranty of
 * MERCHANTABILITY or FITNESS FOR A PARTICULAR PURPOSE.  See the GNU
 * Lesser General Public License for more details.
 *
 * You should have received a copy of the GNU Lesser General Public
 * License along with this library; if not, write to the Free Software
 * Foundation, Inc., 51 Franklin St, Fifth Floor, Boston, MA  02110-1301  USA
 *
 */

#include <Wire.h>    // Library for the initializing the I2C bus.
#include <CurieBLE.h>   // CurieBLE library - pre-installed when Arduino 101 is
selected in Arduino IDE 1.6.7 or later

//I2C Grove Motor Driver defines
//Motor controller defines
#define MotorSpeedSet             0x82
#define PWMFrequenceSet           0x84
#define DirectionSet              0xaa
#define MotorSetA                 0xa1
#define MotorSetB                 0xa5
#define Nothing                   0x01
#define EnableStepper             0x1a
#define UnenableStepper           0x1b
#define Stepernu                  0x1c
#define I2CMotorDriverAdd         0x0f // Set the address of the I2CMotorDriver

// Variables for allocating the Digital I/O pins of the Arduino
//const int resetPin = 2;              // D2 used by Grove Button - reset from crash
state
const int ledPin = 3;                // D3 used by Grove LED socket - showing bluetooth
connected
//const int servoPin = 5;             // D5 used by the servo
//const int ultrasonic_back = 6;  // D6 used by the Back ultrsonic sensor
                                // D7 used by the Back ultrsonic sensor
//const int ultrasonic_front = 8;   // D8 used by the Front ultrasonic sensor
                                // D9 used by the Front ultrasonic sensor
// 0x0f is the I2C address used by Grove Motor Driver board
// 0x3e is an I2C address used by the Grove RGB LCD display (7c>>1)
// 0x62 is an I2C address used by the Grove RGB LCD display (c4>>1)


char vSpeedSet = 50;     // this variable holds the speed setting from the remote
control - default speed is half
char vSpeedLimit = 50;  // this variable holds the limited value of the speed based
on sensor values

char state = 'c';  // initial state is stop
char prev_state = 'c';

BLEPeripheral blePeripheral;  // BLE Peripheral Device (the board you're programming)
// ==== create Nordic Semiconductor BLE UART service =========
// Must use these UUIDs and BLE characteristics
```

```
BLEService uartService = BLEService("6E400001B5A3F393E0A9E50E24DCCA9E");
// create characteristics
BLECharacteristic rxCharacteristic =
BLECharacteristic("6E400002B5A3F393E0A9E50E24DCCA9E", BLEWriteWithoutResponse, 20);
// == TX on central (android app)
BLECharacteristic txCharacteristic =
BLECharacteristic("6E400003B5A3F393E0A9E50E24DCCA9E", BLENotify , 20);    // == RX on
central (android app)


// Setup function - run once at reset or power-on
void setup()  {
  Wire.begin();     // join i2c bus (address optional for master)
  delay(100);
  Serial.begin(115200);
  // initialize the Reset button on D2 and the Blue Grove LED on D3
  pinMode(ledPin, OUTPUT);    // use the LED on pin 3 as an output

  // set advertised local name and service UUID:
  blePeripheral.setLocalName("4WD_RV");    //make unique name
  blePeripheral.setAdvertisedServiceUuid(uartService.uuid());

  // add service and characteristic:
  blePeripheral.addAttribute(uartService);
  blePeripheral.addAttribute(rxCharacteristic);
  blePeripheral.addAttribute(txCharacteristic);

  // assign event handlers for connected, disconnected to peripheral
  blePeripheral.setEventHandler(BLEConnected, blePeripheralConnectHandler);
  blePeripheral.setEventHandler(BLEDisconnected, blePeripheralDisconnectHandler);

  // assign event handler for characteristic
  rxCharacteristic.setEventHandler(BLEWritten, rxCharacteristicWritten);
```

```
  // begin advertising BLE service:
  blePeripheral.begin();
  //Serial.println(("Bluetooth device active, waiting for connections..."));  // for
debugging

  // Initialize the motor controllers
  MotorDirectionSet(0b0000);  //0b0000   stopped
  delay(100);
  MotorSpeedSetAB(vSpeedSet,vSpeedSet); // on a scale of 1 to 100
  delay(100);
}  // end setup

void loop()  {
 /* Key sections of the loop
  *   Check if we are in a "crash" condition
  *   if not crashed:
  *      poll for a new value written into state by the rxCharacteristic
  *      check if there is something new in "state"
  *        if yes call rovercontrol
  *        if no call speedcontrol
  *   if crashed:
```

```cpp
 *     look for a reset button press
 */
   blePeripheral.poll();
   if(state != prev_state) {    // check if the value from the remote control new or
the same as the previous one?
     RoverControl(state);    // function to respond to control state changes -
direction and/or speed
     //Serial.println(char(state));
     prev_state = state;
   }
}

void blePeripheralConnectHandler(BLECentral& central) {    // central connected event
handler
  Serial.print("Connected event, central: ");
  Serial.println(central.address());
  digitalWrite(ledPin, HIGH);
  //Serial.println("LED on");
}

void blePeripheralDisconnectHandler(BLECentral& central) {    // central disconnected
event handler
  Serial.print("Disconnected event, central: ");
  Serial.println(central.address());
  digitalWrite(ledPin, LOW);
  //Serial.println("LED off");
  state = 'c';
  RoverControl(state);                  // stop rover on BLE disconnect - should reduce
runaway rover incidents
  //Serial.println(char(state));
  delay(100);
}

void rxCharacteristicWritten(BLECentral& central, BLECharacteristic& characteristic)
{
  // central wrote new value to characteristic, update LED
  //Serial.print("Characteristic event, written: ");
  if (characteristic.value()) {    // NULL pointer check
    state = *characteristic.value();    // de-reference to get first byte
    //Serial.println(char(state));
  }
}

// Functions to set the 2 DC motor's speed: motorSpeedA: the DC motor A speed; should
be 0~100, motorSpeedB: the DC motor B speed; should be 0~100;
void MotorSpeedSetAB(unsigned char MotorSpeedA , unsigned char MotorSpeedB)  {
  MotorSpeedA=map(MotorSpeedA,0,100,0,255);
  MotorSpeedB=map(MotorSpeedB,0,100,0,255);
  Wire.beginTransmission(I2CMotorDriverAdd); // transmit to device I2CMotorDriverAdd
  Wire.write(MotorSpeedSet);    // set pwm header
  Wire.write(MotorSpeedA);      // send pwma
  Wire.write(MotorSpeedB);      // send pwmb
  Wire.endTransmission();    // stop transmitting
}
```

```cpp
// set the direction of DC motor.
void MotorDirectionSet(unsigned char Direction)  {   //  Adjust the direction of the
motors 0b0000 I4 I3 I2 I1
  Wire.beginTransmission(I2CMotorDriverAdd);    // transmit to device
I2CMotorDriverAdd
  Wire.write(DirectionSet);    // Direction control header
  Wire.write(Direction);     // send direction control information
  Wire.write(Nothing);    // need to send this byte as the third byte(no meaning)
  Wire.endTransmission();    // stop transmitting
}

void RoverControl(char state) {
  /*
   * Respond to changes in direction - forward, left, stop, right, backward from
Bluetooth LE Remote Control
   *
   * With a new "state" from the remote control - check for action to take
   * a - set gear to go forward (drive)
   * b - set gear to turn left and go forward
   * c - set gear to stop (park)
   * d - set gear to turn right and go forward
   * e - set gear to go backward (reverse)
   * 0 - set speed to 0 - not implemented in the remote control
   * 1 - set speed to 25%
   * 2 - set speed to 50%
   * 3 - set speed to 75%
   * 4 - set speed to 100%
   */
  if (state == 'a') {
    MotorDirectionSet(0b1001); }  //0b1001    // If state is equal with letter 'a',
Motors Rotating in the forward direction, rover will go forward
  else if (state == 'b') {
    MotorDirectionSet(0b0001); } //0b0001    // If state is equal with letter 'b',
right motors rotating forward, left motors off, , rover will turn left
  else if (state == 'd') {
    MotorDirectionSet(0b1000); } //0b1000    // If state is equal with letter 'd',
left motors rotating forward, right motors off, rover will turn right
  else if (state == 'c'){
    MotorDirectionSet(0b0000); } //0b0000    // If state is equal with letter 'c',
Motors off - stop the rover
  else if (state == 'e') {
    MotorDirectionSet(0b0110); } //0b0110    // If state is equal with letter 'e',
rover will go backward, both motors rotating backward
  else if (state == '0') {   // Change speed if state is equal from 0 to 4. Values
must be from 0 to 100 - this is mapped to 0 to 255 (PWM) by the MotorSpeedSet
function
    vSpeedSet=0; }
  else if (state == '1') {
    vSpeedSet=25; }
  else if (state == '2') {
    vSpeedSet=50; }
  else if (state == '3') {
    vSpeedSet=75; }
  else if (state == '4') {
    vSpeedSet=100; }
```

```
  delay(10);
  vSpeedLimit = vSpeedSet;
  MotorSpeedSetAB(vSpeedSet,vSpeedSet);    // set motor speed based on changes on a
scale of 1 to 100
}
```

About Author

Ted Humphrey is a tech expert and a programmer who understands python programming language and their application to 21st century problems. Ted has over 12 years of experience writing about latest gadgets and technical appliances. Ted is also a seasoned website developer with 7+ years of experience creating WordPress websites and experienced in using other content management systems, such as Drupal, aside WordPress to create websites. He values making an impact and has a blog where he teaches people the nitty-gritty of Programming Python, developing blogs using WordPress and other content management systems.

Ted holds a Bachelor's degree in Computer science and Engineering from the University of Michigan, USA. He loves pets and he is happily married with two beautiful daughters.

9 798671 121285